Between Blood and Destiny_ A Life Forged by Love, Tested by Fate

QUEEN OF FLOWERS

Published by QUEEN OF FLOWERS, 2025.

While every precaution has been taken in the preparation of this book, the publisher assumes no responsibility for errors or omissions, or for damages resulting from the use of the information contained herein.

BETWEEN BLOOD AND DESTINY_ A LIFE FORGED BY LOVE, TESTED BY FATE

First edition. April 22, 2025.

Copyright © 2025 QUEEN OF FLOWERS.

ISBN: 979-8231398317

Written by QUEEN OF FLOWERS.

"To the One Who Feels Invisible"

If you're holding this book, I want to tell you something first: I see you. I understand the weight of silence, the hunger for kindness, and the longing to be heard. I've been there, and I'm here for you now. For the better part of my life, I was a silent figure. I wore a smile to conceal my pain, served others while starving for kindness, and gave until I forgot the feeling of receiving.

This book isn't just a collection of memories. It's my voice finally breaking through decades of being unheard. It's a hand reaching out to those who've ever felt invisible, forgotten, or broken down by life's cruelty. I penned this memoir not because my story is extraordinary but because it's authentic. I know there are others like me, carrying their pain so silently that even their loved ones can't hear the screams behind the silence.

This memoir is for all of us who have been told to endure, who weep in secret, and who carry our trauma like hidden luggage. It's for my younger version, who only wanted someone to say, "I believe you. You matter." If you find echoes of your struggles in these pages, know this: you are not alone. Your existence is valid. Your emotions are real. And your healing is not a burden-it is your right.

Remember that you're not alone if you find echoes of your experiences in these pages. Your existence is valid, your emotions are real, and your healing is not a burden—it's your right.

This book is my testimony, but it's also an open letter to anyone who wants to be heard. You are seen, you are felt, and you are not forgotten. Your story, your struggles, and your triumphs matter. You matter.

With all my heart,
Queen of Flowers

Introduction

Life is an unpredictable journey—a tapestry woven with love, loss, resilience, and transformation. This book is a testament to that journey, reflecting on the struggles, triumphs, and lessons that shape a person over time. It is more than just a memoir; it is a deeply personal exploration of destiny, faith, and the strength that comes from enduring life's trials.

At the heart of this story lies a woman's pursuit of happiness, belonging, and self-discovery. From cultural and emotional battles to the realities of love, betrayal, and reinvention, this book takes readers through moments of joy and heartbreak that define and reshape identity.

With raw honesty and vivid storytelling, I invite you into my world. In this world, fate and choices intertwine, survival is necessary, and hope is never truly lost, no matter how fragile. This is not just my story. It is a reflection of the resilience we all carry within us.

Turn the page, and walk this journey with me.

Chapter 1:

"The Roots of Resilience:
A Journey Through Unwanted Beginnings"

My story began long before I could understand what it meant to be wanted. I was the baby no one asked for, handed away moments after my first cry echoed into the world.

Fate, however, had other plans. I was placed in the arms of Zenith, a kind woman with soft, wrinkled hands and eyes that longed for a grandchild. She welcomed me into her modest home on the day I was born. She didn't know that her husband, Aaron, had returned from a different direction—with another newborn baby girl.

In a strange twist of destiny, the couple ended the day with two baby girls and one impossible choice.

Their hearts were full, but their pockets were not. They couldn't afford to raise both infants. Eventually, Aaron gave one of us—Lilith—to a trusted neighbour. I stayed behind. Lilith and I would forever be bound by a story neither of us chose.

Zenith and Aaron had no grandchildren then, but they placed their hopes in their daughter, Ella, who had long struggled to conceive. They believed that raising me might bring good fortune. Whether by miracle or myth, it did. Ella soon became pregnant and, ten months later, gave birth to a son.

Ella and her husband, Aspen, eventually moved out to start their own families. I remained with Aaron and Zenith, my first true home.

Our days were simple, but my heart was whole. Aaron had thirteen children with Zenith and ten more from earlier marriages. Our home was always buzzing with laughter, footsteps, and the soft thud of rubber balls being kicked across the dusty yard. I spent my early years running barefoot through rubber plantations, my hair wild in the wind, my voice joining the chorus of childhood freedom. My uncles, barely older than me, were my first playmates. We climbed fruit trees, played traditional games, and shared stolen moments of joy.

But not all memories glowed with innocence.

One afternoon, I jumped from a tree and landed directly onto a cluster of baby chicks, killing them instantly. The silence that followed their tiny cries was deafening. I remember standing frozen, staring at the fragile, lifeless bodies. A wave of guilt washed over me—sharp, cold, and unforgettable. That guilt would shadow me for years, a reminder that even unintentional harm still hurts.

I'd sit beside Aaron in quieter moments as he tinkered with wires and tools, teaching me patience and precision. I loved the sound of metal against metal, the clink of bolts, and how his eyes lit up when I cut aluminium bars. With Zenith, my days were filled with the scent of baking cookies and the rustle of banana leaves as we prepared meals under the big tree behind the house. She was gentle and present. Her presence told me I mattered.

But stability is a fleeting thing.

When I turned five, everything changed. I was sent to live with Ella and Aspen. They were now parents, and I was no longer needed where I had felt most loved.

My biological mother, Fattie—Aspen's sister—and my father, Brendan, never came for me. Not once. I was invisible to them, and the child was erased from memory. In Ella's home, I wasn't a daughter. I was on duty.

The shift was brutal.

Discipline ruled the household like an iron fist in velvet gloves. Ella's methods were cold and calculated. A dish left with soap residue? I had to wash every plate in the cupboard again. Clothes not folded with sharp corners? Redo the entire pile. She had her arsenal—wooden sticks, hangers, a hose pipe—but her favourite was a belt with a rusted metal buckle that left bruises shaped like coins. Her anger didn't flare; it simmered and struck with precision.

One morning, I made her a cup of coffee—watery and weak. She didn't shout. She handed it back and said, "Drink it. All of it." I obeyed. My stomach churned with each sip, but I didn't stop. Hours later, I was sick, curled on the cold tile floor.

Still, I longed to play. To be a child.

Drawing became my only escape. I sketched on anything I could find—walls, books, the back of old receipts. My world became ink and colour. But even that was forbidden. One day, Ella caught me doodling flowers on the back of a calendar. She tore it before me without a word, then locked my pencils away.

I learned early that love, in some houses, came with conditions. And in others, it never came at all.

I wasn't a daughter. I wasn't a sister. I was something in between—an obligation passed around, like an item no one knew what to do with.

Yet, even then, something in me refused to break completely. Maybe it was the memory of Zenith's soft humming in the kitchen or the feel of Aaron's calloused hands guiding mine over a saw. Perhaps it was the baby chicks I couldn't save. But even in the coldest moments, I clung to something unseen. Something that whispered, *you are more than what they say you are.*

I didn't know it then, but those early years were shaping something deep inside me—resilience. Not the loud kind, but the quiet one. The kind that grows in silence, blooms in secret, and waits for its moment to rise.

Chapter 2:

"Bound by Duty, Dreaming of Freedom"

School offered structure—rows of desks, the scribble of chalk on a blackboard, the rhythmic droning of teachers. But once the final bell rang, my world transformed. While others headed off to soccer practice or chatted under the mango trees, I slipped away quietly. My afternoon didn't belong to me—it belonged to duty.

As I opened our creaky gate, the clatter of pots and the sharp scent of garlic sizzling in oil would already greet me from the kitchen. My parents were preparing food to sell at flea markets, often until dusk. Their presence was brief, but their expectations lingered heavily in the air.

I didn't wait to be told what to do. I packed food containers, scrubbed greasy dishes with water that stung my skin, soothed my younger siblings, and swept every corner of the house. The roles blurred—I was a sister, a cleaner, a babysitter, and a child only in name.

When I dashed off to my evening classes, my body ached, my hands were pruney, and my mind felt stretched thin. There were no cartoons to unwind with, skipping ropes or neighbourhood games. Television was a weekend luxury—and even then, just for a moment or two.

Home, for many, is a refuge. For me, it was a terrain of unpredictability. Punishments came quickly and without question. The fault somehow got to me if my brother cried or the floor was dusty. Verbal lashings sliced through the air, leaving invisible wounds that throbbed long after the shouting had stopped.

"You'd have ended up rotting in a ditch if I hadn't picked you up," Ella would sometimes hiss, her words curdled with self-righteousness. *"Nobody wanted you."*

I remember standing outside the dirt path, her voice trailing after me like a siren echoing through the countryside. The neighbours said nothing—only closed their shutters and minds, unwilling to confront her wrath.

But even in this emotional storm, a quiet fire flickered within me: *I had to escape.* I dreamed of living far from her control. And I knew education was my ticket out.

One day, that hope almost slipped from my grasp.

It was exam day—a critical one. I had studied under the dim light of a candle the night before, determined to do well. But as I laced my shoes that morning, Ella stood in the doorway, arms crossed.

"You're not going anywhere," she said coldly. "The dishes are still piled high."

"But the test—" I began, my voice trembling.

"I said no."

I felt helpless. That familiar mix of fear and anger rose in my chest as I stared at the sink. Suddenly, the rumble of a motorbike cut through the tension. My teacher appeared like a miracle in her worn leather jacket and signature confidence.

"Ella," she shouted from the gate, *"she's coming with me."*

For a rare moment, Ella hesitated. I bolted, not daring to look back. That ride to school, wind whipping through my hair, felt like a taste of freedom.

Then there were the school events. On sports days, when the air buzzed with excitement and the smell of fried snacks wafted across the field, I wasn't running with my classmates. I was behind the school canteen counter, helping Ella sell food. I envied the girls in ponytails and sneakers. But I felt strange pride when our prepared food sold out in less than an hour. *At least I was good at something.*

During holidays, there were occasional glimmers of joy. Ella's siblings sometimes cared for Ben and me—Ben, my little shadow, was born ten months after me. We were inseparable. We would escape into nights filled with fireflies and laughter at our aunts' homes. We roasted marshmallows, raced through villages, and made up stories that stretched beyond our reality. Those brief respites reminded me of what childhood was supposed to feel like.

My grandma, the peacemaker, encouraged me to visit our biological parents during Eid or school breaks. They lived just two blocks away from Zenith's house. One visit, though, changed everything.

I was ten. My legs dangled from a rattan chair as I sipped warm tea. I overheard Fattie and Brendan speaking with Ella and Aspen in the adjacent room. Their words tightened something in my chest.

"You should stop sending her to school. What's the point? She's not even your real daughter," Brendan muttered.

"But we're already struggling. We can't afford to keep this up," Ella replied.

They all nodded in agreement. My heart pounded.

They weren't offering help. They weren't grateful. They were planning my failure—my exit from hope.

That day, I left their house in silence. There were no tears, just a deepening disdain. Even as a child, I understood betrayal.

Their voices still haunt me sometimes, reminding me of when I was seen as a burden instead of a blessing. But rather than break me, those words carved out a deeper determination. I would rise—not for them but *despite* them.

Chapter 3:

"Survival, Strength, and Small Victories"

In January 1988, a letter arrived that shimmered with a rare promise. I was thirteen, teetering between childhood and adolescence, when an opportunity presented itself—a chance to attend boarding school. To most, the school wasn't far, but for someone from my background, with barely any transportation and unreliable roads, the journey felt like crossing an ocean. Yet, more than the physical distance, it felt like I was inching closer to a long-held dream: freedom.

To my surprise, Ella and Aspen agreed to send me—reluctantly, yes—but they decided, mostly due to the gentle insistence of my grandparents, Zenith and Aaron. Zenith and Aaron stepped in to support the costs of my schooling. That act of kindness meant everything. I was to spend the next four years of my life—between thirteen and seventeen—within those boarding school walls.

From the moment I arrived, I was swept into a different world. The school buildings were old and worn, smelling of mildew and sun-baked stone. My shoes clicked nervously on the concrete floors as I took it all in—the faded dormitory walls, the high ceilings, the chattering students in mismatched uniforms. But any illusions I had about peace and focus were swiftly broken.

It was a cruel tradition—new students were subjected to relentless bullying by seniors. Though officially banned, the practice continued like an unspoken rule passed down from one batch to the next. We were ordered to perform bizarre tasks: pretending to blow out ceiling lights, washing the seniors' laundry, fetching water in buckets, and even reciting ridiculous lines to earn signatures on a humiliating "checklist." Each signature was a trophy of submission.

But I wasn't the submissive type.

Defiant by nature, I refused to bow to their demands. I didn't collect a single signature. My rebellion earned me a severe punishment—I was forced to stand in the open school field from midnight until morning. The biting wind cut through my thin clothing, the cold seeping into my bones, but I refused to cry. I stood there like a statue, wrapped in stubborn pride and silent suffering. That night broke something in me but also built something else—resilience.

Thankfully, I wasn't entirely alone. Adrian, Zenith's youngest son, quietly supported me by covering part of my monthly fees. Ella and Aspen would visit every two months, bringing two full food baskets. That food did more than fill my belly—it gave me a means to earn. I began selling portions to other students, and the profit helped cover other school expenses. It was a small way of reclaiming some control.

In those days, boarding school fees were $35, and my monthly allowance was a modest $10. I learnt quickly to budget. I stuck to cafeteria food and saved the rest for books, stationery, and the occasional treat. Sometimes, I saved up for months to afford extracurricular activities. The tight budget taught me discipline and strategy—skills that would serve me far beyond those school walls.

Despite the strict rules and hardships, I found comfort in my routine. I was a teenager, and life in the dormitory opened doors to a world I had never known. I had friends. I joined clubs. I participated in after-school activities that made me feel alive. We laughed, played, created—and for the first time in a long time, I felt like I belonged somewhere. Oddly enough, none of my family knew much about this part of my life. My secret joy remained, a space untouched by their criticism or expectations.

One of my most treasured memories from that time was my birthday month. I entered a drawing competition and, to my astonishment, I won. The prize? $180. I'd never held that much money before. My heart leapt with pride and excitement. I bought a cake, ordered food, and threw a mini-party for my friends without a second thought. Laughter echoed through the room that day, and for a fleeting moment, I felt rich—not in money, but joy and companionship.

But good things, as always, carried consequences.

When Ella found out, she was furious. Her voice cracked through the phone, filled with rage. "You're spending money like that while my children have barely enough to eat?" she screamed. I was stunned, holding the receiver in silence. I understood her frustration, but a part of me felt betrayed. Couldn't I be allowed one moment of celebration without guilt?

Despite the blow, I clung to the sweetness of that day. It was mine, and no one could take it from me.

And then there were the mischiefs. Oh, we were far from saints. One afternoon, driven by curiosity and the hunger for freedom, my closest friends and I decided to sneak out of the school. We found a narrow drainage gate, barely noticeable, and squeezed through it with hearts pounding. The adrenaline rush was intoxicating. We boarded a bus to town, giggling like outlaws. But our luck ran out when we realised the warden was on the same bus. The horror! We froze, trying to disappear into the seat cushions, but there was no escape. Though we got into trouble, remembering that escapade still makes me smile. It was one of the boldest—and dumbest—things I ever did at that age.

But not all memories were light-hearted. On 23 December 1989, my beloved grandfather Aaron passed away. I still remember the way grief settled into my chest like a weight I couldn't shake off. He was my champion, my quiet hero. Despite being adopted, he loved me with a rare sincerity. His death left a hollow space in me that no one else could fill.

Still, I pressed on. Pain became my teacher, shaping my thoughts and hardening my resolve. I wore my struggles like invisible armor—never flashy, never spoken of, but always present. Every confrontation with Ella, every unspoken grief, every test passed and task completed—each one was a step towards survival. I was no longer just an unwanted child. I was a survivor, a fighter, a learner of life's cruel, complicated lessons.

During those five years, we were allowed to go home once a month. I usually visited my grandmother Zenith or her sister, who lived nearby. But soon, Ella began discouraging frequent visits. "You're becoming a burden," she said once. "Go home every three months instead." She had her children by then, and life was more complicated. I nodded quietly. I understood.

Long school breaks were never lazy. I took any job I could find—factories, markets, anything. I worked in an electronics factory with my cousin, spent time in a shoe factory alongside Ella, and tried my hands at several other gigs. Whatever I earned went straight to Ella and her family. I didn't complain. I knew my contribution helped. That was enough for me.

I remember one instance in 1990 when I returned to school a week late. My teacher raised an eyebrow and asked where I had been. I gave a vague shrug and said, "I don't know." The truth? No one at home remembered—or had the means—to send me back on time. But I made it. That was what mattered.

One quiet evening, I sat on my dorm bed, a soft breeze wafting through the half-open window. The scent of dust and laundry lingered in the air. My fingers traced the edges of a sketch I was working on—a scene I had drawn countless times—a girl standing free under an open sky. My pencil danced across the paper like a whispered rebellion. It was my therapy. My hope.

The hallway buzzed faintly with the voices of students. I could hear the distant clatter of plates from the dining hall. The smell of rice and fried fish drifted in, grounding me. Five years. Five long years of hardship, joy, discovery, and loss. With my final exams around the corner, I found myself restless. What would come next? Would the world be kinder? Would I finally be free—or would I trade one cage for another?

A knock at the door snapped me out of my thoughts. My roommate peeked in. "You coming for dinner?" she asked.

I nodded, placing my sketchpad down gently.

The light flickered above as I entered the corridor, casting fleeting shadows along the walls. I didn't know then that this would be my last supper in that chapter of life. I knew that the girl who entered boarding school at thirteen was no longer the same. She had grown stronger. Wiser. A little bruised, perhaps—but not broken.

As I walked toward the dining hall, caught between memories and anticipation, I whispered a silent promise to myself: *This is just the beginning.*

Chapter 4:

"From Struggles to Strength: Embracing Independence"

On my last day of exams in December 1992, I received an exciting job offer as a display artist in a shopping mall, which aligned perfectly with my passion and skills. After years of dealing with challenges in boarding school, I thought the real world would be easier. But as I entered my first job, I quickly realised that navigating friendships, workplace politics, and even love came with challenges.

While it may not have been the most prestigious position, it was undoubtedly a better option than remaining at home without purpose. I returned to Zenith's house and began my daily commute to work, although Zaitun expressed concern about the late hours and the inconvenience it placed on her son-in-law, who kindly picked me up from the bus station.

I ventured into a small business, handcrafting T-shirt designs using wax and a traditional batik chanting technique. On one occasion, I nearly set the entire house ablaze, though no one ever found out. The wax I was heating for printing had overheated, and flames had already spread across the wall.

Without hesitation, I grabbed the wax pan, threw a damp towel over the wall to smother the fire, and hurriedly carried the pot to the washroom. I meticulously cleaned every trace of the incident until the stains were completely gone. Only then did the fear set in—an unsettling dread that someone might have witnessed the ordeal.

This experience marked the beginning of my professional journey. Shortly after, Adam, Zenith's son, offered me a more secure living situation, even though it was two hours away from her house. I worked as an assistant graphic designer at a tile factory, making $450 monthly. I carefully tracked my money every payday, giving Adam's wife, Ella, $50 and setting aside $100 for my expenses to continue supporting myself while working.

During my year working there, I formed a close bond with a maintenance guy who was my first love. He introduced me to his parents, but I was pretty young then, and it felt like they didn't fully embrace me. He also came to my house to ask for my hand in marriage, but unfortunately, both families had reservations and accused him of drug use. I felt powerless in that situation, and our relationship gradually faded.

After about a year, I moved on to other opportunities, partly because Ella mentioned that Adam's wife was uncomfortable with my living arrangement. Despite my contributions, it wasn't enough for them. I returned to live with Ella and my uncle Aspen, a positive change. Ella was also going through a difficult period, and her distress was exacerbated when one of her sons was adopted by her sister. Fortunately, I have had no trouble finding work wherever I go. I work at a factory, which has been an enriching experience.

In 1995, I experienced the profound loss of my grandmother, Zenith. Although I can't recall the exact day or month, her impact on my life remains vivid. During her time in the hospital, I took turns with Ella to care for her, which provided a sense of purpose despite the overwhelming loneliness I felt. With both of my beloved grandparents gone, I struggled to cope with the emotional void they left behind. I remember crying deeply, similar to how I felt when I lost my grandfather. It took me several months to process this loss, and I often wished for their support, especially during challenging moments when Ella would act out.

Shortly after that challenging time, I transitioned to a job that ignited my creative passion. I joined a printing company as a graphic designer, earning $400 a month—just enough to cover my transportation and support Ella. At 20, I was learning to navigate the complexities of adulthood, often facing challenges at work that sometimes put me in difficult situations with Ella. Meanwhile, she welcomed another boy into the family, bringing the total to 11 children, which made me the eldest of 12. This new role allowed me to grow and adapt, and I embraced its responsibilities.

A year later, I received an offer from another company that offered double my current wage. Although it was a bit further away and presented some transportation challenges, I decided it was a great opportunity. I began searching for a rental closer to my new workplace and found a suitable three-bedroom house to share with four other roommates, all single professionals working in the same town.

Rather than sharing my plans with Ella and Aspen beforehand, I informed them the day before my new job started. Initially, I intended to book a taxi or a charter car for my commute. Still, my parents offered to arrange transportation for me, perhaps out of concern for my new living situation.

This chapter marks the transition from adolescence to adulthood, where I grappled with distinguishing between colleagues, friends, and mere acquaintances. Although my new workplace was within walking distance from my rented house, my shortcut required a 20-minute walk. One morning, as I passed through a quiet, secluded area, an Indian man suddenly snatched my necklace. I couldn't tell whether he had been following me for some time, waiting for the right moment, or if it was an impulsive act.

Instinctively, I screamed at the top of my lungs, but there was no one around to hear me. Perhaps startled by my cries, the thief panicked and fled, dropping my necklace near the scene. In the struggle, I fell to the ground, scraping my knee and tearing my clothing slightly. Dazed and shaken, I barely registered a passing vehicle that slowed upon noticing my distress.

The driver kindly offered assistance, taking me to the police station to file a report before driving me home. As expected, the authorities showed little interest in pursuing the case—unless a bribe was involved.

This unsettling incident reminded me I was on my own, navigating a world where trust was fragile, and survival demanded resilience. As I immersed myself in my demanding new role, working closely with the owner and sacrificing countless hours without expecting anything in return, I realised that adulthood was not just about financial independence but also emotional endurance. Amidst my struggles, I observed the choices of those around me—my housemates embracing a freer, more reckless existence, some facing heartbreaking consequences, like bringing new life into the world only to be abandoned by the men they trusted.

I watched my housemates venture out, meet men, and seemingly enjoy life with fewer boundaries governing what one might consider a respectable existence. A memory that remains vivid and lingers in my mind involves two of my housemates giving birth alone, without any support. Tragically, their boyfriends refused to take responsibility.

The first of these instances involved a young woman who worked in a factory in a rural area. I recall the events of that day in 1997 as though they were yesterday. Just hours after she gave birth to a baby, I contacted her mother to inform her of the situation. Her mother's response still echoes in my mind—an outpouring of hysterical tears and disbelief that her daughter could have found herself in such a predicament. At the time, society's judgement of children born out of wedlock was harsh and unforgiving, adding yet another layer of pain to an already heartbreaking situation. A mother's love remains unwavering, regardless of her child's age or mistakes. Despite her initial shock, she called me back, her voice steady with resolve, asking me to provide water and food for her daughter as they made their way to collect her and the unexpected grandson. Her care and determination at that moment were a poignant reminder of the enduring bond between a mother and her child.

Not long after that heartbreaking episode, another housemate—my roommate—faced a similar situation. She, too, was a factory worker from a rural background, but she was especially dear to me as we shared countless stories and moments. Thankfully, unlike the first instance, when I only learnt of the situation upon hearing the baby's cries, she confided in me as soon as she realised she had missed her period for two months.

This time, I had the opportunity to seek advice, gather information, and explore options for support. To cut a long story short, one evening, I found myself flagging down a taxi to accompany her to a private clinic she had arranged. The clinic offered her the chance to deliver the baby without charge, but under the condition that the child would remain there, as foster parents had already been arranged. The agreement also stipulated that she would not be given any details about the baby's future or the adoptive family.

At my new workplace, I was fortunate to have a boss who genuinely valued my work ethic and dedication. I remained there for a considerable period, from 1997 to 1999. Each year, when my annual leave came around, I never used it for a holiday. Instead, I devoted it entirely to assisting my mother, Ella, in baking and selling cookies, pastries, and cakes for the festive season.

She would often request that I allocate an entire month's salary to purchase the necessary ingredients, and we would then spend weeks immersed in preparation. There were times when I worked from 7 a.m. until 3 a.m., determined to meet every order, often with the help of my younger siblings. Some requests reached 10,000 pieces, all meticulously crafted by hand without machinery. Exhaustion would sometimes overcome me, and I would drift off to sleep beside the oven, entirely unaware of the sweltering heat. Yet, the hard work was always worthwhile. The profits allowed us to buy new clothes and celebrate the festivities like everyone else. This tradition was not new to me—it was a routine I had been part of since I was six. During this time, two of my younger brothers were getting married, and Ella asked me to take out a loan of $4,000 from my workplace to cover their wedding expenses. She agreed that they would repay me in monthly instalments or as a lump sum. Left with little choice, I complied. However, neither of them ever repaid a single dollar. Now that I work full-time, I feel more like an ATM than a daughter.

It wasn't long before I found myself entangled in a web of rumours, accused of having an affair with him. The drama became unbearable, and one day, I came across an advertisement for a florist position. Without knowing precisely what being a "florist" entailed, I applied for the role, passed the interview, and was offered the job.

I handed in my resignation letter the next day, eager to escape the toxic environment and the tiresome drama. The entire ordeal felt like a scene from a television soap opera—baseless accusations, third parties drawn into the chaos, and unnecessary commotion. The woman knocked on my door repeatedly, often accompanied by her sister, prying into my affairs and involving the neighbours in her misguided vendetta.

Leaving the job and moving to a new place should have marked the end of it, but unfortunately, it didn't. She continued to turn up at my door, creating an atmosphere that left me feeling stifled and overwhelmed. In the end, I resolved to confront the situation directly. I met with my former boss and firmly requested that he put an end to his wife's harassment. I also clarified that I was prepared to escalate the matter by filing a police report if necessary.

Thankfully, the situation began to fade after that. As for what she or others thought of me, I couldn't have cared less. I was born resilient, and no pettiness or unwarranted drama could shake me. Such distractions, while unpleasant, would never break my spirit.

As I walked away from my old life, leaving behind the suffocating rumours, the relentless responsibilities, and the exhausting cycle of giving more than I ever received, I felt a strange mix of emotions. Relief, indeed, but also a lingering ache—from years of being everything to everyone and nothing to myself.

I had spent so long fighting battles that weren't mine to fight, carrying burdens that weren't mine to bear. I had endured accusations, betrayals, and the kind of exhaustion that seeps into the bones. My hands, once calloused from kneading dough and shaping thousands of pastries, now trembled—not from fatigue, but from the uncertainty of what lay ahead.

As I walked away from my old life, leaving behind the suffocating rumours, the relentless responsibilities, and the exhausting cycle of giving more than I ever received, I felt a strange mix of emotions. Relief, indeed, but also a lingering ache—from years of being everything to everyone and nothing to myself.

I had spent so long fighting battles that weren't mine to fight, carrying burdens that weren't mine to bear. My hands, once calloused from kneading dough and shaping thousands of pastries, now trembled—not from fatigue, but from the uncertainty of what lay ahead.

For the first time in my life, I was choosing me. I wasn't running away; I was stepping forward, untangling myself from a web that had held me back for too long. The florist job was an unknown world, but that made it exciting. Flowers didn't gossip, didn't demand, didn't disappoint. They bloomed, and maybe I would, too, in this new chapter.

Still, deep inside, I couldn't silence the minor voice whispering doubts. Would I ever find a place where I truly belonged? Would I ever stop feeling like an ATM, a caregiver, or a convenient problem solver for everyone else's needs? Would I ever be loved, not for what I could provide, but for who I was?

Chapter 5:

"From Heartbreak to Horizons Unknown"

The morning light filtered through the sheer curtains, casting delicate patterns on the wooden floor. A soft breeze carried the faint scent of jasmine—a fragrance that once symbolised hope but now felt like a distant memory. She stretched, feeling the stiffness in her muscles, remnants of a restless night filled with unresolved thoughts.

As her fingers traced the embroidered edges of the bedspread, a ritual that had always grounded her, she reminded herself: *Today is a new beginning.*

A new place. A new chapter. A fresh start.

When an opportunity arose at a prestigious golf resort in 2000, I took it as a sign. This was my chance to move forward, to leave behind the ghosts of my past and create a life filled with possibilities. I had spent years surviving—now, I wanted to live.

It didn't take long for me to be noticed at my new workplace. My petite stature, boundless energy, and ability to juggle multiple tasks efficiently made an impression. In less than three months, I had carved a place for myself. I also met a man—one who would soon change everything.

It didn't take long for me to be noticed in my new workplace. My petite stature, boundless energy, quick learning, and ability to complete a considerable amount of work in a short time all garnered my attention. In less than three months, I fell in love with a man, and it was a highly complex relationship. We formed an instant connection. Our evenings were filled with laughter—simple yet joyful moments spent at cafés, burger joints, or within the walls of our staff accommodation. It felt effortless as if life had finally aligned in my favour.

Then, one day, the ground beneath me crumbled.

He came to me, his face etched with distress, and confessed a truth I could barely comprehend. He had been coerced into marriage.

The year was 1999, and in his culture, traditions were still deeply rooted in the past. He told me how it happened, his voice heavy with regret.

"She invited me to her house, making sure no one else was home. The moment I stepped inside, the religious authorities and neighbours ambushed me. According

to our laws, a man and woman found alone together must either marry or face severe punishment. I had no choice but to agree."

I could barely breathe. My mind reeled, my heart shattered. This wasn't just a betrayal—it was a deliberate trap, orchestrated to ensure he would never be mine.

I needed answers. Desperation drove me to confront the woman who had taken him away from me. My fingers trembled as I typed the message, asking her why she had done it and knowingly shattered what we had.

Her reply was chilling in its simplicity.

"I love him too, and I don't want him to marry you. That's why I trapped him. And it worked."

Those words burned into my mind, an inescapable reminder of the cruel reality I had to accept.

I was utterly devastated. In a fit of frustration, I sent a message to the other woman, asking why she had acted in such a manner, knowing full well that he and I were in a relationship at the time. Her response was chillingly simple: "I love him too, and I don't want him to marry you. That's why I trapped him, and it worked." I am speechless. Nevertheless, I refused to let the situation dampen my spirit. I carried on with life, working and living as though nothing had happened, though deep down, the hurt lingered.

I refused to let heartbreak define me. I buried myself in work, forcing normalcy into my routine. Each day, I went through the motions—smiling when expected, performing my duties flawlessly, pretending that I wasn't breaking inside.

Then, another opportunity came knocking—a position at another prestigious golf resort in the same city. I accepted, determined to start over once again.

But this time, I had learned my lesson. No more love. No more distractions.

I threw myself into work and was online when I wasn't working. In the Yahoo and MSN Messenger era, virtual connections could quickly turn into real-life friendships. I joined various online groups, engaging in meetups and social activities that filled the void in my life.

One group, in particular, became my escape—Lovers' Corner.

Lovers' Corner wasn't just a chat group but a community. We hosted monthly meetups filled with lively discussions and exciting activities. One particular event stood out—one I had the privilege of hosting at my workplace.

Ironically, the manager who once issued me a formal warning for using the company computer for personal matters later praised me for bringing in business through that event.

Life had its contradictions, and I was learning to navigate them.

Despite my growing recognition at work, fate had other plans for me.

An opportunity arose—one that promised double my current salary. My decision was driven not by ambition but by necessity.

My sixth sibling had secured a place at a private university, and between my salary, Aspen's, and our other siblings' earnings, we still couldn't cover the tuition fees. Ella insisted that I take out a bank loan in my name, sure that my employment at a prestigious resort would ensure approval.

She was right. The loan was approved. And from that moment on, my entire salary was swallowed by repayments.

With nothing left to survive on, I had no choice. I accepted the offer in Brunei Darussalam.

I have always been impulsive, but this decision felt different. It was a leap into the unknown, driven by obligation rather than desire.

I told my parents the day before my departure. I never formally resigned from my job.

My reasoning was simple—*if Brunei didn't suit me, I would return within a week.*

But as the plane ascended, I felt something crack inside me. Tears streamed down my face, soaking into my white polo shirt—my best clothing. I didn't know why I was crying.

Was it sorrow? Relief? A mix of both?

I had no answer.

Surprisingly, I adapted quickly. The two-bedroom apartment provided to me was comfortable, and my roommate—a woman of mixed Bruneian and English heritage—was warm and welcoming.

Despite my department director's lukewarm impression of me, the salary was enticing enough to keep me there for nearly a year.

I formed a small but close-knit circle of friends—my roommate, a fellow Malaysian, her boyfriend, and a few acquaintances from the culinary team.

Food had a way of forging bonds. Whenever the culinary team needed floral decorations for last-minute events, they would appear in my florist's room bearing cheesecake or other sweet bribes. I pretended to resist, but we all knew I never could.

While in Brunei, I seized every opportunity to explore—road trips to Miri and Limbang, ferry rides to Labuan and Kota Kinabalu. Each journey brought a fleeting sense of freedom.

But Brunei, for all its wealth and oil-fueled luxury, was a country wrapped in silence.

By 6 p.m., the streets were empty. Shops closed. Life slowed to a halt.

It was peaceful, but peace wasn't what I craved.

I longed for movement, excitement, something more than this quiet existence.

Then, one afternoon, the phone rang.

I answered absentmindedly, expecting a call from my Thai colleague, but instead, a voice on the other end posed a question that would change everything.

"Would you be interested in relocating to Dubai?"

I said yes without hesitation.

I didn't know where Dubai was. I didn't care.

From that moment on, my heart no longer belonged to Brunei.

I would sneak into the hotel lobby, using the internet café—three Brunei dollars per hour—to exchange emails with my prospective employer. My body remained in Brunei, but my soul had already left.

And when the day finally came to leave, I felt no sadness, no regret.

Only exhilaration.

I was stepping into the unknown.

And I felt truly alive for the first time in a long while.

Chapter 6:

"Dubai Before the Skyline:
My Early Days in an Unknown Land"

On January 22, 2002, just days after my 27th birthday, I landed in Dubai, eager to embrace a new beginning. When I stepped off the plane, an unforgiving wave of dry, desert heat greeted me. The city stretched before me, a barren landscape of endless sand, with only a handful of buildings scattered across the horizon. It was a world in the making—raw, rugged, and unpolished.

I had always prided myself on adaptability, but nothing could have prepared me for the overwhelming culture shock that awaited. Once again, the language barriers, loud arguments, and unfamiliar customs reminded me that I was an outsider. The air felt charged with a different rhythm of life—unfiltered, intense, and sometimes jarring.

Dubai was a collision of cultures, and not all encounters were welcoming. Some people spoke with fiery passion, their words laced with emotion, while others disregarded social decorum altogether—gesturing with their feet instead of their hands, staring unapologetically, or openly displaying prejudices. Some dismissed Asians outright, insisting on dealing only with Westerners or those in managerial positions.

But I had never been one to cower. I stood my ground, carrying myself like a fighter—ready, if necessary, to deliver a metaphorical knockout.

At work, things were manageable, at least on the surface. The job itself felt familiar, unlike the hotels I had worked at back home and in Brunei. But beneath the polished exteriors was an undercurrent of superiority—an unspoken competition between nationalities, each eager to assert their dominance. As one of only three people from my country, I explained where I was from more often than I cared to. Eventually, I stopped trying. My focus remained on my work, and I kept my social interactions to a minimum.

Weekends became my escape, offering small pockets of familiarity in an unfamiliar world. I found solace in Satwa and Jumeirah Village, where I would meet fellow countrymen, exchanging stories and laughter over simple meals. Sometimes, my Thai colleague and I would walk to Lamcy Plaza, a short kilometer away, and then wander into Karama in search of budget-friendly finds.

Dubai was unrecognisable in those days compared to what it is now. There were no towering skyscrapers, no sprawling malls—just a city still finding its shape. The roads were narrow, traffic sparse, and getting lost was nearly impossible. Back home, people barely knew where Dubai was, often confusing it with Mumbai. Without smartphones or Google Maps, I relied on an old encyclopedia, carefully pointing out my location to friends and family.

Despite the challenges, the year passed swiftly. My purpose remained clear: I was here to work, to support my family. By the end of the year, I had repaid my sister's education loan in full. I sent my entire salary home every month, keeping only $50 for myself. With free meals, accommodation, and transport provided by the company, I needed little else. My world shrank to the four walls of my shared room, where I forged a deep friendship with my Burmese roommate. Our late-night conversations revolved around work, cultural nuances, and the peculiarities of life far from home.

Then, everything changed on the eve of Ramadan.

Just as I was about to drift into sleep, a firm voice called out,

"Wake up and pray."

I was alone. And yet, I obeyed.

Rising to perform my prayers, I wept uncontrollably. It had been nearly a decade since I had last prayed—back in my boarding school days. Something had stirred inside me, something I couldn't explain. When I finished, I reached for my only Kashmiri scarf, wrapped it around myself, put on a long dress, and stepped out for dinner.

The following day, I went to work transformed. Gone were the short trousers, fitted T-shirts, and flip-flops. Instead, I dressed modestly—long trousers, a long-sleeved shirt, proper shoes, and a headscarf. It was Ramadan, and no one questioned my change in appearance. But this wasn't about tradition. It wasn't about society. I had heard a voice, and I had answered.

The real battle began when Ramadan ended.

My manager, a South African woman, accused me of violating company policy, claiming that my hijab was against grooming standards. I was summoned by HR and ordered to remove it. Calmly, I pointed out the hypocrisy—if rules were enforced, why was the HR manager allowed excessive jewellery? Why was my manager's bright nail polish overlooked? I documented every inconsistency and submitted a formal letter to HR headquarters.

Word spread quickly, and to my surprise, support arrived from unexpected places. Several Muslim colleagues of Arab descent rallied around me, advising me to seek guidance from the Islamic Department in Dubai. Their suggestion? Take the matter to the Royal Advisor's office.

So we did.

I still remember how the Royal Secretary slammed his hand on the table in fury. "Foreigners in this country must do their jobs without changing the local laws," he declared, his voice sharp with authority. He promised immediate action.

By the time we stepped outside, the news had already travelled. The Chairman of my company—a European man with stark white hair—had received a call from the Royal Office. The rebuke was swift and severe. The next day, I was summoned for a one-on-one discussion with the vice chairman, a distinguished gentleman with a paralysed hand. To my surprise, he wasn't there to reprimand me. Instead, he wanted to understand. Why did I choose to wear the hijab?

Not everyone shared his curiosity. Some colleagues—people I once called friends—turned against me, telling me I was in the wrong industry to practice my faith. Their words stung, but I refused to let them define me.

Days later, I received another unexpected visitor—the Corporate HR Director. Unlike the formal meetings before, this was unofficial. He assured me he wanted to help.

And then, the unexpected happened.

Instead of being dismissed, I was transferred. No longer a florist, I was reassigned as an HR passport controller at the company's headquarters. The very people who had sought to remove me now found themselves speechless.

As I settled into my new role, the world outside was shifting. The Iraq War erupted, sending ripples of instability across the region. Tourism declined, hotel occupancy plummeted to 10%, and the company encouraged employees to take unused annual leave. I seized the opportunity to visit my family. Despite Dubai being over 1,800 kilometres from the conflict, they had grown increasingly worried.

During my time at home, an offer came—one that changed everything. My former supervisor, who had long admired my work, offered me a new opportunity—this time, not in Dubai but in Malaysia. A wholesale flower company needed someone with my expertise to help supply the Dubai market.

When my leave ended, I returned to Dubai—not to resume work, but to resign.

When I handed in my resignation letter, I felt an unexpected weight settle over me. Relief, nostalgia, apprehension—they all intertwined in that moment. Dubai had been more than just a workplace. It had been my battleground, classroom, prison, and sanctuary.

As I walked through the city one last time, I inhaled deeply—the scent of oud lingering in the evening air, the distant hum of the call to prayer merging with the rhythm of traffic. The skyline, once dust and sand, had begun to rise.

I wasn't just leaving a job.

I was living a life I had built from nothing. A life that had made me who I was.

But the world was calling again.

And I had never been one to ignore its call.

Chapter 7:

"A Ruthless Welcome"

Two months later, I was in Malaysia, embarking on a new venture. I had barely settled in when I was thrust into a terrifying ordeal—one that would forever shape my perception of safety.

It happened in broad daylight, just a week after my arrival. We were unpacking a shipment from Denmark, eagerly unwrapping crystal vases, and preparing flowers and foliage for dispatch to Dubai. The villa we rented—a two-story space that doubled as an office and accommodation—was bustling with activity. Then, in a matter of seconds, the atmosphere shifted.

The door burst open. Five men stormed in, their faces hardened, their hands gripping sharp, glinting cleavers. Before we could react, they had us restrained—our wrists bound tightly with rough ropes. I could feel the coarse fibres digging into my skin as panic settled in my chest.

When it was my turn, they ran out of rope. Without hesitation, one grabbed a printer cable and wound it around my wrists. I held my breath as another yanked the gold necklace from my neck, then tore my bracelet from my wrist. My grandmother's ring—one of my most treasured possessions—was still on my finger, but with quick thinking, I slipped it under the table, praying they wouldn't notice.

They took everything—cash, phones, cameras, office computers, and other equipment—before vanishing as quickly as they had arrived.

For several minutes, we remained frozen, the air thick with fear. Then, slowly, we began helping each other untie the knots. My manager, who had miraculously managed to conceal his phone, dialled the police. We also had an illegal worker from Bangladesh with us, and in the chaos, we urged him to leave before the authorities arrived.

Nearly an hour later, the police finally showed up. I had expected urgency, thoroughness, and maybe even reassurance. Instead, they moved with routine detachment, treating our ordeal like a minor inconvenience. There were no forensic teams, no fingerprint analysis—none of the meticulous procedures I had seen in films or TV dramas. Just a few notes scribbled on a notepad and a half-hearted remark:

"This area has never had a robbery before."

I wasn't sure if that was meant to comfort or unsettle me.

At the station, I sat with an officer to give my statement and help create a photofit of the culprits. But with CCTV cameras scarce then, the chances of catching them felt slim. I walked out of the station with a hollow sense of reality—this was Malaysia, where I had come to chase a new beginning. And yet, within days, it had already thrown me into survival mode.

Despite the unsettling start, Malaysia became essential to my professional growth. I worked closely with my former supervisor from Dubai, who became my manager. The industry was fast-paced, and I immersed myself in learning the intricacies of business and the creative field.

But soon, I struggled with something I hadn't anticipated—his personal life bleeding into work. His relationship with his boyfriend constantly interfered with business, turning our workplace into an emotional battleground. Arguments erupted without warning, tension lingered in the air, and they would abandon their responsibilities when things got terrible, leaving me to pick up the pieces.

At first, I told myself to endure it. But as time went on, I realised that the instability was affecting my mental well-being. I was exhausted—not just from the workload but from the emotional chaos.

I needed a way out.

When I finally landed another job outside the capital, I took it without hesitation. The daily train commute—nearly an hour from Kuala Lumpur city centre—wasn't ideal, but at least it was a fresh start. A week into my new role, a colleague mentioned they had a house for rent near my workplace. Moving there meant cutting my commute to just ten minutes by train. Without a second thought, I agreed.

The job was routine—working in a small flower shop inside a hotel. There was little room for creativity, but it gave me stability. I stayed for less than a year before moving on to a competitor, relocating back to Kuala Lumpur with a double salary.

But with the financial gain came another sacrifice—balance.

My new employer was an acquaintance from my previous workplace, the same place where I had been robbed. I oversaw operations for a company specialising in decorations and linen rentals for hotels and event organisers this time. The work was relentless, and my living conditions were less than ideal. My room was no more than a cramped space, just large enough to fit a single bed. But rent-free accommodation meant saving money, so I accepted it as part of the trade-off.

Then, one afternoon, another unexpected confrontation with the police occurred.

My team and I had just crossed the road for lunch when a patrol car pulled beside us. The officers stepped out, their eyes scanning us before demanding identification. At first, I didn't think much of it—until I noticed the tension in my crew's body language.

Unbeknownst to me, some of them were illegal Bangladeshi workers carrying falsified passports. I only realised it that moment, watching their silent exchanges. The officers weren't concerned about law enforcement; they were looking for an easy payday. And they had found one.

A tense negotiation followed, culminating in me withdrawing cash from an ATM to pay them off. It wasn't right, but I had no choice. My team couldn't afford to be detained because I had a significant event to handle. I played my cards carefully; as expected, the officers took the money and left.

It was yet another reminder of the reality I was living in—a constant tightrope walk between survival and ambition.

My career continued to push forward. I met clients, developed creative concepts, and moved from one hotel to another, orchestrating grand events. But behind the success was exhaustion.

There were nights when I had nowhere to sleep, choosing to rest in the warehouse because there was no other option. The loneliness was suffocating. With no family or close friends around, I felt like a ghost moving through the city—visible to everyone yet unseen.

The months dragged on, and the thrill of creating extravagant events began to fade. Instead of feeling accomplished, I felt depleted. What was the point of all this if I had no personal life, no stability, no place I could truly call home?

The final straw came when I caught my reflection in a mirror one evening—worn, drained, and lost. This wasn't the life I had envisioned for myself. I was merely surviving, not living.

Then, an opportunity in Oman surfaced.

The thought of a new beginning in a different country, working at a five-star luxury resort, felt like a breath of fresh air after a storm. Could it be different this time? Would I finally find a place where I belonged?

I didn't overthink it. I packed my belongings, booked my ticket, and boarded the flight.

Leaving Malaysia wasn't just about changing jobs but breaking free from a cycle that had drained me. The long hours, the lack of personal space, the weight of responsibilities that weren't mine to carry—it had all been too much. I had learned valuable lessons, but at a cost, I was no longer willing to pay.

As the plane took off, I gazed out the window, hopeful and hesitant. I didn't know what awaited me in Oman.

But I knew one thing: I was chasing more than a career this time.

I was chasing a life worth living.

Chapter 8:

"Mountains, Memories, and Molestation:
A Chapter of Oman"

Arriving in Oman felt like stepping into a different world—a world of serenity, sophistication, and an unmistakable sense of grandeur. The palace hotel where I was to work stood like a jewel against an endless blue sky. It was magnificent, and I felt excitement for the first time in a long while. Perhaps, this was exactly where I needed to be.

Oman was unlike anywhere I had been before. When I arrived, I was greeted by a landscape that felt untouched by time—majestic mountains, endless desert sands, and the deep blue sea stretching into the horizon. The air was different, carrying the scent of frankincense and a sense of calm I hadn't known in years.

From the beginning, I felt something I hadn't felt in a long time—a sense of home. The people were warm, their hospitality unmatched. My colleagues, both local and international, became my new family. Here, we didn't just work together; we lived together. Even on our days off, there was always something to do—cooking together, exploring the markets, or setting off road trips into the mountains. Out of pure generosity, some locals would invite us to their villages, showing us the beauty of their homeland—picking fresh grapes, figs, and pomegranates straight from the farms, discovering ancient forts, or taking boat rides deep into the sea.

I found solace in nature and adventure. Mountain climbing became my escape. Sometimes, I would go with friends and other times; I would go alone. Standing at the top, looking down at the vast landscape, I felt something I hadn't in years—peace.

But Oman, like every place, was not without its shadows. For a while, Oman felt like the escape I had longed for. The warmth of the people, the breathtaking landscapes, and the strong sense of community were a world apart from the chaos I had left behind in Malaysia. But just as I started to let my guard down, reality struck.

It happened at work, within the very place where I had started to feel at home. One of my colleagues, someone I had to interact with daily, violated my trust in the worst way possible. I was molested. The shock, the disgust, and the sheer anger coursed through me, but I refused to stay silent. I took the matter straight to HR, determined to hold him accountable.

But the response I received was something I will never forget.

Instead of support, I was met with dismissiveness, indifference, and excuses. The HR officer, a woman, looked at me coldly and said, *"Try to understand. He has a family."* As if that justified what he had done. As if my dignity and my safety meant nothing in comparison.

What about me? Did I not deserve respect? Did I not have my value as a person? The anger boiled within me, but I knew I was fighting a battle I could not win. Here, in this place, silence was expected. Justice was not a given but a privilege reserved for a few.

I walked out of that office with a bitterness I could not shake. It wasn't just about the man who had harmed me—it was about the system that protected him. I had arrived in Oman believing in its beauty, kindness, and sense of community. But at that moment, a part of me started to resent it. I knew not all Omanis were the same, but the experience had planted something dark inside me—a distrust I couldn't ignore.

And yet, life had its way of challenging my beliefs.

Not long after this incident, I stumbled into an unexpected connection. A local man, someone I had met on my flight from Malaysia to Oman, reappeared in my life.

I had completely forgotten about him until one day while reorganising my room, I came across a scrap of paper with his number scribbled. I hesitated for a moment, then thought—*why not?* Maybe having someone by my side would make me feel safer. Maybe it would keep other men from seeing me as a target.

I reached out. And so, it began.

Our meetings were simple—dinners, casual walks through the markets, quiet conversations in the car. He lived only a few kilometres from my staff compound, making it easy for us to see each other. For the first time since arriving, I allowed myself to relax in someone's company.

But happiness, it seemed, was always fleeting in my life.

One evening, he broke the news.

He was engaged—to his cousin. It was a family arrangement, something deeply rooted in their traditions. There was no way out for him. As a government employee, he risked losing his privileges—his job, his standing in society—if he married a foreigner.

We both sat silently, the weight of reality pressing down on us. It felt unfair, cruel even. Just as I started to feel something real, it was ripped away. But maybe, deep down, we had always known. Maybe we were never meant to last.

There were tears, quiet acceptance, and finally, the understanding that we had no choice but to move on.

Despite everything, I couldn't bring myself to hate Oman. It had given me so much—a thriving career, friendships that felt like family, and a work-life balance I had never experienced. For the first time in years, I wasn't drowning in exhaustion or consumed by survival. I had space to breathe, explore, and simply exist without the constant weight of uncertainty.

But even amid contentment, loneliness lingered like a shadow I couldn't shake. I had people around me, but no one genuinely close to my heart. I had laughter, but no one to share my deepest fears or quiet moments with. I was surrounded by warmth and camaraderie at work, but I went home alone at the end of the day. The silence of my room felt deafening.

I often found myself staring out at the mountains, letting the vastness of the landscape swallow my thoughts. I had accomplished so much—I had built a career, carved a place for myself in a foreign land, and gained respect in my field. But what did it all mean if I had no one to come home to?

I was happy, but I was sad.

Oman had given me stability, adventure, and purpose. But it had also reminded me of what I lacked—love, true companionship, and a sense of belonging beyond the workplace. I had left Malaysia searching for something more, and in many ways, I had found it. Yet, the void in my heart remained, growing deeper each day.

I didn't know what the future held. I didn't know if I would ever find the connection I longed for. But as I stood beneath the endless Omani sky, I made a silent promise to myself—I would keep searching. I would keep moving forward, even if I had to do it alone.

Chapter 9:

"The Rise and Realities of Entrepreneurship"

It started with a phone call. Somewhere between 2006 and 2008, while I was still working in Oman, my former supervisor from Dubai—who had since become a manager—reached out to me with an unexpected proposal. He wanted to start a company and asked me to be his business partner.

At first, I hesitated. Running a business was exciting, but it felt like stepping into the unknown. Yet, something inside me told me this was an opportunity I couldn't ignore. So, I agreed. I invested $10,000 to kickstart the venture, as did my two partners—him and his boyfriend, whom I had once worked with in Malaysia. Ironically, Malaysia held some of my worst memories, like the time I was robbed with a cleaver in my neighbourhood. But those fears didn't haunt me anymore. I had grown.

A few months later, the hotel I worked at in Oman announced a significant refurbishment project that would take approximately two years. Staff were being transferred to other properties within the same hotel chain. I tried applying for a cross-transfer to Singapore and Australia, but my hijab became a silent barrier. While many of my colleagues secured placements in different parts of the world, I was rejected. It was frustrating, but in hindsight, it was a blessing in disguise. This was the sign I needed—I chose to resign and pour my energy into the new business.

The early days were nothing like the glossy images of entrepreneurship I had seen in magazines or movies. There were no luxury cars or designer suits—just sheer survival. The company specialises in events, floral arrangements, and decorations, catering to five-star hotels and high-end venues. The potential was there, but the first three months were brutal.

We had drained our savings to fund the business from scratch, paying suppliers, staff, and operational costs. When everything was settled, I checked my pocket and found only $2 left. I still remember that moment vividly—the silence between us, the unspoken fear in our eyes. We had taken a leap of faith, but the reality of entrepreneurship was far from glamorous.

That night, I prayed harder than ever, surrendering everything to the Almighty, the provider, and the merciful. The very next day, we received our first significant payment from a hotel—a $15,000 contract. That was our turning point. We rolled the money back into the business, carefully managing every cent, ensuring that we could sustain ourselves while continuing to grow.

From that moment, things took off. The demand for our services grew, and soon, I found myself constantly on the move, managing projects between Kuala Lumpur and Penang. It was exhausting. I was always running, working long hours under the sun, losing weight, my skin darkening from the constant exposure. But the business was thriving, and that was all that mattered.

In Kuala Lumpur, we rented a three-bedroom apartment and converted it into our office. This time, we chose a strategically located guarded property in the Golden Triangle area, with easy access to all modes of transportation and the luxury hotels we worked with. Efficiency was key.

We became even more prominent in Penang by renting a massive two-story villa. The ground floor and yard became our office and warehouse, while the upper floor served as our accommodation. Every inch of space had a purpose.

I had no time for love, no time for companionship. My life revolved around the business and the business alone. Relationships require time, patience, and emotional energy—none of which I had to spare. I wasn't unhappy, but I wasn't entirely fulfilled either.

The only drama in my life came from my business partners—the couple who never stopped fighting. Their conflicts split into our work, and whenever tensions exploded, they would disappear, leaving me to handle everything alone. I didn't sign up for this, but I had no choice. I had to carry it on my shoulders if I wanted the company to survive.

And so, I did.

Despite the challenges, I told myself this was temporary. One day, the hard work would pay off, and I wouldn't have to do it all alone. I just had to keep going.

The first year was a battlefield. We struggled financially, experimented with ideas, and fumbled through strategies. None of us had prior business experience, and every decision felt like a high-stakes gamble. But while we were young and inexperienced as entrepreneurs, we were far from amateurs in our industry.

Each of us brought something valuable to the table—my strength was in operations and creativity, ensuring our floral arrangements and event designs were nothing short of spectacular. My friend handled business development, creativity, and public relations, securing clients and forging connections in the luxury market. His boyfriend, the numbers guy, managed our finances and also played a role in public relations. It was a well-balanced team—at least on paper.

The reality was much messier.

A year into the business, things started looking up. We had established ourselves in the five-star hotel circuit, gained the trust of high-end clients, and secured long-term contracts. The financial strain was easing, and for the first time, we had room to breathe. To take things to the next level, we planned a business trip—part work, part pleasure—to China, Hong Kong, and Macau.

Back then, online suppliers weren't as widespread as they are today. If we wanted quality materials at competitive prices, we had to go straight to the source. That meant visiting factories, negotiating deals in person, and hauling samples ourselves. It was exhausting—mentally, physically, and emotionally. The days were long, filled with endless meetings and factory visits. The nights weren't much better—sleep was scarce, and stress was constant.

But it was worth it.

We built direct relationships with suppliers, cut costs, and expanded our inventory. This trip proved that we were evolving—not just surviving but growing.

Yet, for all our business progress, things behind the scenes were unravelling fast.

The drama between my two partners, once just an occasional annoyance, had escalated into a constant storm. Their relationship bled into work, making every decision, every meeting, a battleground. Arguments turned into screaming matches. Silent treatments lasted for days. And every time they fought, one would vanish, leaving me to pick up the pieces.

It felt like déjà vu—just like when I first arrived in Malaysia years ago. Mixing business with personal relationships had never worked in my world. It drained me and left me empty, and no amount of success could make up for the emotional exhaustion.

I tried to push through. I told myself it would pass. That they would grow up, that things would get better. But they didn't.

And one day, I had enough.

I sat in our Kuala Lumpur office, staring at the city skyline, the weight of my exhaustion pressing down on me. The business was thriving, and we were stable financially. But at what cost?

I was tired of being the mediator, of the emotional chaos, and of feeling like I was running a marathon with no finish line. The work itself had never scared me. I could handle all the late nights, early mornings, and endless responsibilities. But the toxicity? That was something I wasn't willing to endure anymore.

So, I made my decision.

I sold my shares, packed my bags, and booked my flight back to Oman.

The timing couldn't have been more perfect. The hotel I had once worked at was completing its refurbishment, and they were looking to rehire staff. It felt like the universe was giving me a reset button, a chance to start over.

As my plane soared above the city, leaving behind the business, the fights, and the stress, I didn't feel regret. I felt relief.

Chapter 10:

"Waves of Change:
Finding Stability in an Unstable Time"

Returning to Oman felt like slipping back into a well-worn, familiar rhythm. The sights, the sounds, and even the scent of frankincense in the air brought a sense of comfort. This time, though, things were different. The hotel was gearing up to reopen after its long refurbishment, and my role naturally expanded beyond floral arrangements and décor. I volunteered to assist in the office, handling paperwork, organizing inventory, and updating the progress reports. The sense of teamwork was strong—we weren't just colleagues; we were a family. Many of my coworkers, especially within my department, were related by blood, hailing from nearby villages, their bonds forming an unbreakable network of trust and support.

Outside of work, I immersed myself in new passions. My love for mountain climbing remained steadfast, but now I also found exhilaration in kayaking and photography. Oman's landscapes provided the perfect backdrop—the golden hues of the mountains at sunset, the turquoise waters of the Gulf, and the chaotic yet beautiful energy of Muttrah Souq. After work, I often wandered through the souq, breathing in the scent of spices and oud, or took evening strolls along Barr Al Jissah beach, where the waves whispered against the shore. Some weekends were spent on a boat, riding the restless waters of the Omani Gulf, while longer holidays led to new mountain explorations. In the winter, we hiked deep into the rocky terrain, encountering mountain goats with their thick, fluffy coats, their curious eyes watching us from a distance.

Life here was good, but something felt stagnant. I thrived on inspiration, on the constant rush of learning and creating, yet my career had hit a plateau. The hotel remained prestigious, but I needed fresh challenges. Every so often, I escaped to Dubai, hopping from one luxury hotel to another, absorbing new trends and gathering ideas. The city had transformed dramatically—skyscrapers piercing the sky, new hotels sprouting up like mushrooms after the rain. It was no longer the quiet, almost unfamiliar place I had arrived in years ago; now, it pulsed with energy, ambition, and a relentless drive forward.

Then, towards the end of 2010, the atmosphere in Oman shifted. The ripples of the Arab Spring reached its shores. Street protests broke out, including outside the very hotel where I worked. Many of my colleagues joined in—not out of desperation or suffering but swept up in the momentum of the movement. They were well-paid, and their government took good care of them, but the collective energy of change was intoxicating. It didn't reach the same level of chaos as other Arab nations, but the unease settled deep in my bones. Oman had always been a sanctuary of peace, but now, uncertainty lingered in the air.

A year later, the inevitable news arrived—an international hotel chain would take over management. The announcement sent a wave of mixed emotions through us. Sadness and anxiety, but also an understanding that nothing lasts forever. Change was the nature of life—just as good things eventually end, so do the bad. The new management arrived like a storm, making sweeping changes. They brought a different corporate culture and a different pace, and while some embraced it, I found myself struggling to adapt. The warmth that had made this place feel like home started to fade.

Sensing the shift, I made a decision. Six months before my contract ended, I quietly began searching for new opportunities. Some of the previous management had already left the country, and a few took my CV with them, promising to reach out once they secured their own new roles. I didn't know what the next step would be, but one thing was clear—Oman had given me so much, but it was time to look beyond its shores once again. With the departure of the old management, everything at work had changed. New faces, new leadership styles, new standards—it felt like starting from scratch in a place I had once called home. The sense of belonging I had built over the years was fading, and the unfamiliar energy in the workplace left me unsettled. I waited for the calls from those who had taken my CV, hoping for a promising opportunity, but none came. Instead, the unexpected happened—I received an offer from Kuwait.

I had never considered Kuwait before. The country had never been on my list, and the things I had heard about it didn't exactly inspire excitement. But my options were limited. I knew I couldn't stay in Oman much longer—I despised the new working culture, and my heart no longer felt connected to this place. The only way forward was to take a leap of faith. So, I accepted the offer.

While waiting for my hiring documents, I returned to Kuala Lumpur, using the time wisely. I had been saving up for years, and I decided it was finally time to buy a place of my own. The apartment wasn't brand new, but it was in good condition, only five years old. Property prices were still affordable at the time, so I seized the opportunity. I spent the next few months handling legal paperwork, dealing with banks, and finalizing the purchase, all while doing part-time work with my former company—the one I had once been a shareholder of.

But three months passed, and my savings had dwindled. Between purchasing the apartment and covering my expenses, I found myself with almost nothing left. When the hiring documents finally arrived, I faced a harsh reality—I had no money to start fresh in Kuwait, no safety net, and no backup plan.

But giving up was never an option.

I did what I had to do—I sold all my gold jewellery, every last piece. I had no regrets. I was starting over, at zero. There were no debts, loans, or borrowed money—just me, my determination, and the unknown future waiting for me.

In January 2012, I landed in Kuwait.

The moment I stepped out of the airport, the air felt different—thick with dust, the kind that clung to your skin and settled into every crevice. Kuwait was still healing from the scars of war. It had been 20 years since the invasion, yet much of the city remained undeveloped, frozen in time. The streets had an air of neglect, and many places had a rustic, almost abandoned feel.

I quickly realized that this was not going to be an easy place to live. American soldiers roamed the city, a lingering presence from the war era. And the way men treated Asian women—especially Southeast Asians—was appalling. It reminded me of my early days in Dubai, but this was worse. The stares weren't just curious; they were invasive and predatory. Walking down the street felt like walking through a battlefield, eyes scanning me like an MRI machine, stripping away every layer of dignity.

Some men didn't even try to hide their intentions. I saw them unzipping their pants in broad daylight, their eyes locked onto any woman who walked past. It was a different level of disrespect, a reminder that in this part of the world, some still saw Asian women as nothing more than objects.

The worst was when my Korean flatmate was chased by a group of young men—laughing, taunting her as they advanced, their zippers down, their intentions clear. She ran, her fear palpable, but in Kuwait, there was no guarantee of safety. No one intervened. No one cared.

And I wondered—had I made a mistake coming here?

The new hotel where I worked was locally owned, run by one of the seven wealthiest families in Kuwait. It was a stark contrast to my past experiences in Oman and Dubai. Kuwait, though a small nation, boasted the highest-valued currency in the world. Yet, for all its wealth, warmth and kindness seemed to be in short supply.

Hospitality, a fundamental part of the industry, was strangely absent here. I quickly observed how the elite treated those they deemed beneath them—with an air of entitlement, raised voices, and a complete disregard for basic respect. Shouting at lower-level employees seemed like second nature to them, a habit ingrained in their culture. It grated on my nerves, but I learned to tune it out. I refused to let their arrogance define my experience.

Instead, I focused on creating my own space, my own circle of trust.

I reconnected with a few acquaintances from Oman and slowly built friendships with colleagues from different departments and varying positions within the hotel. Unlike in Dubai or Oman, where socializing felt effortless, here, it required more intention. I found that food was the perfect bridge. I often brought home-cooked meals for lunch, and just as the old saying goes—*food brings people together*.

Even the General Manager, a man who had once lived in Southeast Asia, became enamoured with my cooking. He eagerly requested that I share my recipes with the hotel's chefs and even suggested incorporating my dishes into the hotel menu. It was a small but meaningful victory—one of the few positive experiences in an otherwise challenging environment.

As the days passed, I adapted to the rhythm of my new life. I learned the unspoken rules of the workplace, mastered the art of selective hearing when tempers flared, and found ways to keep my days smooth and easy. But beyond work, Kuwait itself remained unfamiliar, uninviting.

One afternoon, restless from the monotony of our routine, my Korean flatmate and I decided to explore the city. We ventured out into the blinding heat of summer, determined to see what Kuwait had to offer.

It was a mistake.

The temperature soared close to 60°C, an unbearable furnace of dry, suffocating heat. With every step, the soles of our shoes felt as if they might melt into the pavement. The air burned against our skin, our bodies protesting against the extreme conditions. Within minutes, dizziness crept in. It felt as though the world around us was warping, the oppressive heat draining every ounce of energy we had.

But we refused to turn back.

Determined, we pushed forward, navigating the city's souqs, wandering through mosques, and browsing the vast yet lifeless shopping malls. There was no vibrancy, no charm, just a dusty city that seemed to exist in a state of perpetual haze.

By the time we returned to our flat, exhaustion hit us like a tidal wave. We collapsed onto the couch, drenched in sweat, gulping down ice-cold water in desperate relief. My Korean flatmate let out a breathless laugh.

"We survived."

I nodded, feeling the same mix of triumph and absurdity.

Life in Kuwait wasn't easy. It lacked the vibrancy of Malaysia, the adventure of Oman, and the opportunity of Dubai. But I had survived worse, and I knew I would survive this too.

Because no matter where I went, I always found a way.

Maybe it was because I had already convinced myself that Kuwait was nothing more than a temporary stop, a place to endure rather than embrace, but no matter how much time passed, I struggled to find anything truly fulfilling here. Unlike Oman, where warmth and adventure were woven into daily life, Kuwait felt... empty.

Yes, I made good friends among my colleagues, and yes, I settled into the routine of work, but beyond that? There was nothing worth holding onto. The energy of the place was harsh—men shouted rather than spoke, hospitality was a foreign concept to many, and respect was something only a select few received. The arrogance of wealth loomed over every interaction, and the treatment of Asian workers, especially women, was something I would never grow accustomed to. I kept my head high and refused to be intimidated, but deep inside, I knew—this was not my place.

Still, amidst the loneliness, I fell in love again.

At first, it felt different. I allowed myself to believe that maybe this time, things would be real, that maybe this was the moment I would finally stop running. I let my guard down, thinking that after everything I had been through, I had found someone who saw me—not as an outsider, not as a convenience, but as a person worth choosing.

I was wrong.

The truth unravelled slowly, piece by piece until it shattered everything I had foolishly built in my mind. He was getting married. Back in his home country, an entire life awaited him—a life that never included me. And worse, I wasn't the only one he had strung along. There was another woman, just like me, just as disposable in his world of selfish desires.

How foolish I felt.

How naive.

How utterly stupid to believe.

I had survived betrayals before, but this one cut deeper, not because of him—but because of me. How could I have let this happen again? How could I have allowed my heart to hope, even for a moment, in a place that had given me nothing but reasons to leave?

I withdrew, disgusted with myself for allowing such a man into my life. But if Kuwait had taught me anything, it was that there was no point in dwelling on pain—not in a city that did nothing to ease it.

So, I shut it away, buried it where no one could see, and began reaching out to the people who had once made me feel like I belonged.

Oman.

Not the place itself, but the people who had once made it home.

When I got back in touch with my former management team—now settled in Dubai—I felt something I hadn't felt in a long time.

Hope.

The moment I heard their voices, I knew.

This was it.

The light at the end of the tunnel.

And this time, I wouldn't let it slip away.

Even before I landed in Dubai, I had already woven myself into the fabric of my new workplace.

I had been working remotely, designing proposals, submitting ideas, and building a relationship with my future employer. Not for the money—there was no extra pay—but because it filled me with purpose and excitement. Perhaps it was because I already knew who I would be working with. Perhaps it was because I had already walked these streets before, felt the pulse of this city, and knew that Dubai was different.

I was ready.

In early August 2012, I took annual leave to visit home. But before heading back, I planned a brief stopover in Dubai—just for a few days.

A chance to say hello to the new venture, the new team, and the old faces I had worked with back in Oman.

Before stepping onto my one-way flight out of Kuwait, I had one last thing to do.

I sat down, opened my laptop, and began to type. An email.

Not to a colleague.

Not to a friend.

But to the owner of the hotel I had worked for—the very place where I had spent months enduring coldness, arrogance, and the absence of true hospitality.

The words flowed effortlessly.

I told her everything—the unfair treatment, the unwelcoming culture, and the hollow leadership that ran the hotel without warmth or care.

I wasn't writing for closure.

I wasn't writing because I thought anything would change.

I was writing because my voice mattered—even if no one cared to listen.

I hit send. And just like that, I was done.

Chapter 11:

"Blinded by Love: A Warning Ignored"

As I stepped off the plane and into the heat of Dubai's summer, a rush of energy coursed through me. The towering skyline, the hum of ambition in the air—everything about this city felt alive, and so did I.

It was as if I had been reborn.

After months of enduring Kuwait's oppression, of feeling trapped in a city that never felt like mine, Dubai was like a breath of fresh air.

But before I could officially join, I had to wait for my working visa. I had learned from past experiences not to sit idle, so I flew home and took on part-time jobs at weddings and events.

Being home should have been comforting, a chance to pause, to breathe. But it wasn't.

Something inside me always pulled me away.

I tried to settle in, to convince myself that this time, I could stay—but the feeling never lasted.

Home was too stagnant, too predictable.

I longed for movement, for the unknown, for a life beyond the four walls of familiarity.

Was it determination?

Restlessness?

Or just the unshakable truth that I was meant to live differently?

Maybe I wasn't running.

Maybe I was searching.

Searching for something bigger.

September 11, 2012.

I arrived in Dubai not as a visitor, but as a resident once again.

But this time, I set no expectations.

I wasn't chasing dreams, nor was I seeking permanence.

I had learned that life was unpredictable, that no place was truly permanent, and that I was, by nature, a nomad—moving from place to place, following opportunities, following hope.

Dubai was just another stop on my journey.

And yet... something in my heart told me—

This was only the beginning.

Dubai, a city that felt both familiar and foreign, welcomed me once again. But this time, it was different. The company I joined was a Canadian hotel chain, and within its workforce was a melting pot of nationalities—Africans, Arabs, South Asians, Russians, Chinese, Filipinas, and Indonesians. And as always, I was the odd one out.

Nobody ever guessed my nationality correctly. Filipino? Indonesian? Nepali? Indian? Burmese? Egyptian? Moroccan? The guesses were endless, but never right. Maybe it was my features, my accent, or the way I dressed. Or perhaps their understanding of the world was simply too limited.

It didn't matter. I was used to it.

What did matter was that, despite my hard work and efficiency, I had once again become the subject of whispers and gossip.

I had always worked faster than my own thoughts, relying on my guts and instincts. And in a place like Dubai, being exceptional often made you a target.

My salary was higher than others of the same rank, which was enough to turn admiration into resentment.

One Egyptian manager, unable to digest this fact, spread rumours—claiming that I was earning as much as he was, despite not being in a managerial role. The word "wasta" surfaced often.

> *Wasta*—the Arabic term for connections, influence, and favouritism.
> A system where who you know often matters more than what you know.

I didn't care. Let them talk. Their words wouldn't change my paycheck or the value of my work.

While others found comfort in complaining, I immersed myself in learning.

Since the hotel had yet to open, my role wasn't clearly defined. Some days, I worked closely with the admin team. On other days, I was assigned to clean storerooms, organize paperwork, and even scrub toilets.

Nothing was beneath me. Every task was a lesson. Every moment was an opportunity.

I welcomed it all, knowing that versatility was my greatest asset.
Among the many people I worked with, one stood out. I called him PIA, an Egyptian who had secured his position through "wasta"—a favour from the Aussie Director of Housekeeping.
Despite our vastly different backgrounds, we clicked instantly.

We weren't just colleagues; we became friends.

After long days of unpredictable work, we'd hang out, watching football matches, movies in staff accommodation, or organizing short trips to the city.
It was a small window of freedom before the storm—because once the hotel opened, we knew what was coming:
16-hour workdays. Little sleep. Even less pay.
So, we made the most of our pre-opening days, carving out moments of joy before reality hit.
During this time, my passion for photography reached new heights.
I wanted to hone my skills, so I asked my colleagues if they'd model for me. They agreed, and soon, we planned a day out in old Dubai—Bur Dubai and Deira.

Abra rides across the creek. Strolling through vibrant souks. The scent of spices in the air. The weight of history in every alleyway.

It was a perfect day, except for one problem—none of them had money.
With salaries yet to be deposited, they were broke. So, I became the lender, ensuring that our little adventure wouldn't be ruined by empty wallets.

We ate cheap street food. Took photos in the relentless 45°C heat.
Laughed like carefree children.

It was one of those rare days when life felt simple again.
Looking back, that chapter of my life was bittersweet.

I had proven myself in the workplace, yet success came at the cost of envy and isolation.

I had built friendships, yet I knew they were temporary, fleeting moments in a transient city.

I had pushed myself to learn, yet I constantly wondered—what was I truly chasing?

Dubai was not just a place of work. It was a battlefield—one where only the strongest survived.

And I was determined to be one of them.

My relationship with PIA started as a playful friendship—we laughed, teased, argued, and spent endless hours doing silly things together. But soon, friendship turned into something more.

We became a couple.

And just like that, I unknowingly boarded a rollercoaster ride—one that would take me through love, frustration, and countless battles of pride.

One afternoon, we decided to go to a nearby mall. It was a simple plan, nothing extraordinary.

"Let's take the bus," PIA suggested casually.

I hesitated. I wasn't used to public buses—I usually took taxis or private cars. But out of respect for him, I agreed.

"Fine. Let's do it your way," I said, trying to hide my reluctance.

In Dubai, you don't pay for a bus ride with cash—you tap a transportation card on the scanner, and that's it. I tapped mine, heard the beep, and walked in.

But then, PIA decided to play the hero.

"Give me your card," he insisted with a smug smile. *"I'll tap it for you again."*

I rolled my eyes but handed him my card. If he wanted to show off, fine.

A few minutes into the ride, a ticket inspector boarded the bus. I barely paid attention—after all, I had tapped my card. What could go wrong?

Then, the inspector approached me.

"Madam, your card doesn't show a valid scan."

I frowned. "What do you mean? I tapped it."

"No," he said sternly, pointing at his device. *"It says you tapped in, then immediately tapped out. You haven't paid."*

I turned to PIA, realization sinking in.

"This is because of you!" I whispered harshly. *"You tapped my card twice—once for in and once for out!"*

PIA froze, his confidence suddenly gone.

I refused to pay the fine, even though it wasn't my fault. But before I could argue further, the inspector confiscated my Emirates ID (EID)—something no resident wants to happen.

Now, I was furious.

Not only did I have to get off the bus in the middle of nowhere, but I also had to deal with the embarrassment of being fined for something I didn't even do.

"You wanted to be the hero? Then fix this," I snapped at PIA.

But then, he disappeared.
I called him. No answer.
I called again. Still nothing.
I clenched my fists. Where the hell had he gone?
At this point, I could have paid the fine myself, but I refused. PIA wanted to handle it, so I let him handle it.
Eventually, after what felt like an eternity, he came back. Without a word, he handed me my EID.

"Where the hell were you?" I demanded.
"I withdrew money and paid the fine," he muttered.

I wanted to explode.

> "So you abandoned me in the middle of all this? You caused the problem, disappeared, and now you're acting like you did something noble?"

I had enough. I waved for a taxi.

> "Get in," I said coldly.

PIA crossed his arms, his ego bruised.

> "No. I'll take the bus."

Fine. Let him. I left without him, my anger simmering.
That night, I texted him.

> "Where are you?"

No reply.
Later, when I saw him, he had a sulking expression, acting like I had caused the problem.
I could have apologized to keep the peace. But why should I?

> He had messed up. I had suffered for it. And now, he wanted to act like the victim?

We fought, but it didn't last long. Our arguments never did.
Still, something changed that day.
I realized love isn't just about romance. It's about how someone handles conflict, how they take responsibility, how they treat you when things go wrong.
And for the first time, I wondered—was I truly with the right person?
From that bus ride fiasco, a seed of doubt had already been planted in my mind. Was PIA truly the right person for me? Or was I simply forcing myself to believe in something that wasn't meant to be?
But I convinced myself to stay.

I told myself, " Nobody is perfect." People change, and relationships take effort. Maybe I just needed to be more patient.

I tried to rationalise his flaws, thinking love was about tolerance and that with enough time, we would find our rhythm.
But then, his birthday happened.
It was meant to be a small, simple celebration—just cake, snacks, and a few close friends.
Everything started fine—laughs, casual conversations, and music playing softly in the background. But then, PIA decided he wanted to "have fun."
Without warning, he grabbed the birthday cake—a beautiful, chocolate-covered masterpiece—and smashed it onto our faces.
At first, we laughed, thinking it was a harmless joke.
But he didn't stop there.
He took handfuls of cream and frosting, smearing it on the walls, the floor—anywhere but the plates. Then, with a sudden burst of energy, he grabbed one of his friends and threw him into the pool.
The laughter died down.
Everyone was furious, but we forced smiles, pretending it was all in good fun—after all, it was his birthday. But in reality, we were done with his antics.
From that moment on, whenever someone had a birthday, we did not tell him—just to avoid another disaster.
And still, I stayed.
The real breaking point came later, behind closed doors—a moment that shook me to my core.
One day, PIA borrowed my laptop. I didn't think much of it at the time.
But then, he found something—an old conversation between me and the Egyptian guy from Kuwait. The same one who had lied to me, used me, and was engaged to another woman.
I had long since moved on, but that didn't matter to PIA.

"So, you're still talking to him?" his voice was sharp, accusing.

"It's nothing," I tried to explain. *"It was a casual conversation. Nothing romantic."*

But he didn't listen.

We fought. He was angry. I was exhausted.

The stress weighed on me, so I did the only thing I knew how to do—I took a sleeping pill to calm myself down and drift off.

But PIA had other plans.

I had just begun to feel the effects of the pill—my body growing heavy, my thoughts blurring into the edges of sleep—when suddenly,

BANG!

The door slammed open.

Before I could react, PIA was on top of me, his hands around my throat.

"What the hell are you doing?" he hissed.

I tried to fight back, but my body was weak and drowsy from the pill.

"Spit it out! SPIT IT OUT!" he demanded, his grip tightening.

I gasped for air, panic surging through my veins.

Did he think I was trying to kill myself? Did he think I was running away from him?

I tried to mumble a response, but I could barely breathe.

The room spun, and for the first time, I felt real fear—fear of the person I thought I loved.

Eventually, he let go, stepping back with wild eyes, his chest heaving.

I coughed, rubbing my sore neck, too shaken to speak.

That night, I lay in bed, staring at the ceiling, asking myself one simple question:

Why am I still here?

I knew this was a red flag. A massive, screaming, flashing red flag.

And yet, six months later, when PIA proposed,

I said yes.

Why?

Because I wanted to believe in him.
Because he promised to change.
Because he told me he would take care of me.
Because I was 38 years old, and society had conditioned me to believe that a woman my age shouldn't still be single.

> *"I will take care of you," he told me, his voice softer than before. "We will grow old together."*

I wanted to believe those words. So I closed my eyes to the warnings. And I walked straight into the storm.

Chapter 12:

"A Celebration Without Joy"

A month before the wedding, I sat alone with my thoughts, contemplating calling it off. A voice inside me whispered that something wasn't right, that I was making a mistake. But somehow, despite the doubt lingering in my chest, we pushed through.

There was no party, no celebration—just an agreement that we would have one later when we visited our hometowns. Perhaps that was just an excuse to avoid acknowledging the emptiness I already felt. The entire process had taken three months, drowning in endless paperwork, waiting for documentation from his country and mine. Maybe that delay should have been my sign. Maybe that uncertainty wasn't just about logistics—it was about us.

On June 24, 2013, we woke up early, got dressed, and prepared to register our marriage in court. But in typical fashion, what should have been a meaningful day turned into another fight.

The reason? Transportation.

He hadn't made any arrangements, so at the last minute, I called a private taxi—something similar to Uber or Careem, though unregistered and illegal in Dubai at the time. Instead of being grateful, he was furious.

"You always take charge like a man," he snapped, his voice laced with irritation.

I stared at him, speechless. Was this really happening? On our wedding day?

I was exhausted—emotionally and mentally. No one was beside me. No family, no friends. Just him, sulking over his bruised ego. But instead of crying, I just felt... drained. Too tired to fight, too fed up to care. Perhaps I had already given up before it even began.

Two of our colleagues, sworn to secrecy, agreed to be our witnesses. They smiled and tried to make small talk, but I could tell they sensed the tension.

There was no celebration afterwards. No dinner, no flowers, no photographs to capture the moment.

I had secretly longed for something—even just a quiet dinner, a small cake, something to acknowledge this life-changing moment. But he refused to spend a single penny. Not even for our wedding.

Once the ceremony was over, we left by train.

Because he didn't want to pay for a taxi.

Because even on the day he was supposed to be my husband, he still refused to make me feel special.

I looked down at my dress—not new, not special. Just an old dress I had owned since 2004. Nearly a decade old, worn and faded, like the love that never really existed between us. He didn't want me to dress up, didn't want me to wear anything that made me look like a bride.

And I?

I didn't fight it.

Looking back, I know I was blind and foolish. The moment I saw the truth, I should have walked away, but I accepted my fate, trusting that the Almighty had a plan greater than mine.

Because He is the Knower.

He is the Planner.

And He had something waiting for me that I had yet to see.

Three months after our civil marriage, we travelled to Egypt so I could meet his family for the first time. Neither of our families truly supported our decision, but at this point, they had little choice but to accept it.

The moment we landed, I was greeted by his two brothers and father. I had no idea what to expect, but the drive from the airport quickly filled me with unease. Cairo was unlike anything I had ever seen. The traffic was chaotic, with cars tailgating so closely that I was certain they would crash at any moment. There were no pedestrian crossings, no rules—just people darting across the streets, weaving through speeding vehicles like they had no fear of death.

I gasped as a man jumped right in front of our car, narrowly avoiding being hit. My husband and his brothers burst into laughter, finding my terror amusing.

"You'll get used to it," one of them smirked.

But I didn't. Even years later, they still laughed at how I clung to my seat during that drive.

At some point, I noticed the driver seemed unsure of where he was going. I watched as we veered away from the city lights, deeper into darkness. The paved roads disappeared, replaced by dirt paths and vast fields stretching endlessly into the night. My stomach twisted with unease.

"Where are we?" I asked, my voice tight.

No one answered immediately. Then, with a chuckle, one of his brothers muttered something in Arabic.

I didn't understand their words, but I understood the tone—casual amusement as if it were just a joke. But nothing felt funny to me.

For the first time since arriving, I felt real fear. What if something happened to me here? Who would know? Who would care?

Eventually, after what felt like hours, we found our way back to the main road, and the tension in my chest loosened. Looking back, I realize it was a ridiculous situation, almost comedic. But in that moment, trapped in the dark on an unknown road, with no control over what happened next—it was anything but funny.

The next morning, I was barely awake when my mother-in-law began making calls, one after another. She was summoning her siblings, cousins, extended family—every one—to meet me and celebrate our marriage.

I should have felt honoured. Instead, I felt like an exhibit.

Before I could even process what was happening, his cousins and aunts whisked me away to go wedding dress shopping—something I never wanted. But out of respect for their culture and expectations, I forced myself to go along with the charade.

Shop after shop. Dress after dress.

I felt suffocated. Every gown looked the same to me—ornate, heavy, and overwhelmingly extravagant. I wanted nothing to do with it, but I swallowed my discomfort and smiled.

At one point, I noticed my mother-in-law discreetly slipping off her wedding ring, preparing to pawn it to buy my dress. The realization stunned me.

The dress wasn't expensive—barely 200 dirhams when converted to UAE currency. I reached into my bag, ready to pay, but she immediately refused. She wouldn't even call her son to ask for money.

The ego. The same pride I saw in her son, I now saw in her.

I stood there, watching her willingly part with something sentimental just to keep up appearances, and I wondered: Was this what my life would be like? A constant battle between pride and practicality?

The night before the wedding, everything exploded.

I discovered that he had invited his ex-lover to our wedding. Not only that, but he had planned to meet her before the event and even deleted her number from his phone to hide it from me.

That was it.

I refused to stay silent. The anger surged through me, and before I knew it, we were in a heated argument.

"If she means that much to you, why don't you marry her instead?" I snapped.

His face darkened. His jaw clenched.

He looked at me—not with regret, not with guilt—but with frustration. Like I was the problem.

"I don't have to explain myself to you," he muttered.

He was ready to call off the entire wedding reception, and honestly? I wouldn't have minded.

But then, his mother intervened.

I watched as she pleaded with him, urging him to respect the marriage and go through with the celebration.

By that time, some of his close family members had already arrived. The house was filled with people—talking, preparing, expecting a wedding.

And once again, I was trapped.

That night, I lay awake, staring at the ceiling, feeling the weight of my choices.

I thought about how, just a month before, I had wanted to call off the wedding but convinced myself to push through.

I thought about the fights, the secrets, the warnings.

I thought about how no one—not my family, not his—truly supported us.

And yet, I went through with it anyway.

Because deep down, I had told myself that maybe, just maybe, things would get better.

But as I listened to the distant sounds of his family laughing in the next room, I already knew—

This wasn't the beginning of happiness.

It was the beginning of something else entirely.

The day after the wedding, everything felt oddly familiar, yet strangely new. We spent the next week sightseeing, making our way through Cairo's bustling streets, and visiting the eldest members of the family. Each day brought a whirlwind of new experiences, but the underlying chaos of the city never faded. We hopped on and off public transport—buses, trains, shared vans—riding through the heart of Cairo's frenzy. The city, alive at all hours, never seemed to sleep. For me, it was a constant hum of activity, the noise, the crowds, the relentless movement.

Cairo seemed to breathe in dust and noise, its pollution hanging in the air like an invisible shroud. The streets were crowded with people shouting, desperately trying to sell their goods, weaving between the traffic with practiced ease. Some areas demanded long, exhausting walks to find transport, the roads full of holes, mud, garbage, and the unmistakable scent of horse and donkey faeces—a stench that clung to everything. I tried not to breathe it in, but it was impossible to escape.

Yet, through it all, I kept my composure. I never complained, even when the smells overwhelmed me or when the exhaustion from the chaotic transport left my body aching. Instead, I smiled or nodded, pretending it didn't bother me.

The family—especially the aunts and cousins—were impossible to ignore. They had a knack for prying into matters I had no intention of discussing. "How's your life now?" they would ask, their voices full of curiosity and warmth. Their questions were relentless, personal, and intrusive, even as they dug into the smallest details of my life. I pretended not to understand their words, although I could hear the laughter in their voices, their eyes gleaming with a knowing light. It was a culture of closeness, of sharing, and of knowing everything about each other. In their eyes, it was normal. In mine, it felt like a constant test of my privacy and my boundaries.

I kept wondering:

Was this what family meant here? A tight-knit group where personal space didn't exist but love was always in abundance? It was a strange kind of warmth from being surrounded by people who never stopped talking or asking.

Though I was uncomfortable at times, I couldn't help but admire their connection—a bond forged through years of shared stories, laughter, and, yes, the occasional intrusion.

As the days passed, I began to reflect on how much had changed in my life. I had gone from being a quiet observer to being thrust into a completely foreign world. But there was growth in it—an understanding that not all love had to be expressed in silence. And sometimes, even when the noise and chaos of life felt overwhelming, there was a strange beauty in being swept up by it.

We had the opportunity to visit family in Alexandria, which meant a nearly three-hour train ride. The journey was long, and every station seemed to drag on forever as the train made countless stops. The atmosphere inside was a chaotic mix of sounds and smells. Youths would jump from the moving train at each stop, avoiding the fare collector with a mix of bravado and reckless abandon. I watched, heart racing, as they leapt from the train, thinking for a moment that something was wrong. My mind immediately imagined an emergency—a fire or another disaster. The men around us, unfazed, continued to smoke their cigarettes, the air thick with the acrid scent of tobacco. I tried to hold my breath, but it was impossible to escape the fumes, and I could feel the suffocating weight of the smoke in my chest. They spat out the windows, its stains marking the train's path from Cairo to Alexandria.

The madness unfolding before me disoriented my thoughts. *Is this normal?* I wondered, feeling trapped in a world that seemed so far removed from the calm I had known.

When we finally reached Alexandria, our journey wasn't over. We had to walk, yet again, for what felt like miles from the train station to the family home. Every step reminded me of how far we'd come—from the fast-paced, noisy life of Cairo to the slower, but no less chaotic, streets of Alexandria. The family, ever mindful of their expenses, chose to walk instead of taking a cab, saving every penny. It was a constant reminder of how little could be spent in a place where frugality was prized above comfort.

Once we arrived, they offered us a meal—one that was familiar yet distant. It was the same food my parents would prepare back in Cairo, a simple but hearty meal, full of love but also tinged with the reality of a life shaped by careful budgeting and sacrifice. Despite the exhaustion of the journey, I felt a strange sense of belonging in their small home, a warmth that wasn't about the food on the table but about the bonds we were beginning to form.

Though far from the private honeymoon I had imagined, this trip marked a turning point. It was the beginning of a new reality—it wasn't just the two of us whenever we went anywhere. The family was always there, tagging along, part of every outing. At first, it felt overwhelming, but I began to understand its necessity as the days passed. Without them and our family's constant presence, perhaps we would have slipped into more tension and misunderstandings. In a way, their presence was a buffer to avoid the kind of daily arguments that could easily have arisen in the silence of solitude.

Still, I couldn't help but think back to what I had imagined a honeymoon would be—just the two of us, alone, discovering new things together. But perhaps this was how we would discover each other, not in quiet moments of isolation but in the constant presence of those who had been part of his life for so long. And there was a different kind of intimacy—one shaped not by seclusion but by shared experience.

At this moment, I felt like a stranger in my skin. Nothing about this felt like *my* life. I was no longer the woman I knew myself to be; I was simply a reflection of the expectations and desires of those around me. Every decision, every plan, seemed to be made for me by them. I had become an observer in my own story, a passive participant in a foreign life.

We moved from one family gathering to another, from one meal to the next, as though we were following a script already written for us. The food was always what they wanted, and the activities were based on their schedules. There was no room for me to express what I truly wanted, no space to carve out a path that was mine.

It dawned on me slowly—this wasn't my wedding. This wasn't my honeymoon. It had become his family's wedding and honeymoon. The joy, the celebrations, and the intimacy I had imagined for myself all seemed to have been swallowed by the whirlwind of his family's expectations. And though I smiled and nodded, a quiet sadness settled in my chest. It wasn't anger or frustration—it was a calm realisation that the life I had envisioned for myself in this new chapter had been shaped entirely by others.

One evening, as we sat in a small café, surrounded by the noise of the bustling streets, I turned to him and said, "This isn't what I thought it would be." The words felt strange as they left my lips, a mix of longing and resignation. He looked at me, his eyes soft with understanding. "I know," he said quietly. "But this is how we do things here. Family is everything."

I sat back, the city's noise filling the space between us, and thought about his words. The family was everything. But in that moment, I realised that in their world, there was no room for the quiet intimacy of a honeymoon, no space for the privacy that I had imagined. I had come to a crossroads where my desires were being slowly swallowed by a culture of togetherness—one that, at least for now, didn't leave much room for me to be alone with him, to find our rhythm.

But there was something I couldn't ignore. As I sat there, watching families come and go, sharing meals and stories, I understood that perhaps this wasn't entirely a loss. Maybe there was value in this closeness, in the constant presence of family. Perhaps, in time, I would learn to navigate this new world, balancing my desires with the deep ties that bound his family together.

After spending nearly a month in Cairo, we finally returned to Dubai. The familiar skyline greeted me as we landed, starkly contrasting Egypt's chaotic, dust-filled streets. I should have felt relieved to be back, but something inside me felt... off. It was like I had stepped out of one world and into another, but neither felt like home.

On our first morning back at work, we brought a box of chocolates to share with the department during the morning briefing—a small gesture, a quiet celebration of our new life together. The room fell silent as we unwrapped the sweets and passed them around. Confused glances were exchanged, and murmurs passed between colleagues.

"Wait, what?" one of them finally blurted out. *"You two got married?"*

I forced a small smile, nodding as I sipped my coffee.

The Aussie Housekeeping Director, a sharp-eyed woman with years of experience behind her, called me into her office later that day. I stepped inside, expecting a casual chat, maybe a work update. Instead, she leaned forward, her expression unusually serious.

"Are you okay?" she asked, her voice measured.

I blinked. *"Yes... why?"*

She hesitated for a moment, choosing her words carefully. *"I lived in Egypt for quite some time. I know how things are there. And I know how women from your country grow up, the kind of life they're used to."*

A strange unease settled in my stomach. Was I that transparent? Did I *look* like someone who needed saving?

She softened her tone. *"If you ever need help—anything at all—just let me know."*

I could feel her watching me, studying my reaction. I took a steady breath, forcing my expression into something unreadable.

"Thank you," I said politely, *"but everything is fine."*

I wasn't sure if she believed me. I wasn't even sure if *I* believed me. But I knew one thing—I didn't like people prying into my life. The last thing I wanted was pity or concern.

I wondered why her words unsettled me so much as I left her office. Maybe because deep down, a part of me *wasn't* delicate. Perhaps because she had seen something in me that I hadn't yet admitted to myself.

Chapter 13:

"Bound by Vows, Chained by Silence"

My new life wasn't easy. I had spent years being independent, making my own choices without a second thought. I followed my instincts, trusted my gut, and did what felt right. But now, everything has changed.

Every decision—big or small—was no longer mine alone to make. I had to ask for permission for things I had once taken for granted: what to eat, what to wear, who to talk to, who to meet, and even who to greet. Each time, I reminded myself that this was part of marriage and being a good wife. But the more I adjusted, the more I felt like I was losing pieces of myself.

We lived in a tiny studio apartment in staff accommodation, barely big enough for a single bed, a small sofa, and a TV. There was no kitchen, no space to cook. The walls felt like they were closing in on me. There was no corner to retreat to, no place to escape and be by myself.

I had imagined marriage to be a partnership, a union where two people supported each other. But in reality, it felt like I had stepped into a role I didn't fully understand. I told myself to be patient, adjust, and embrace this new life. But the truth was, I felt suffocated.

And then, there was the issue of money. I had worked hard for my independence, earned my salary, and always had control over my finances. But now, every dirham I earned had to be accounted for. I couldn't spend without his approval. At first, I told myself it was just a form of respect, a way to keep transparency in marriage. But over time, I realised how trapped it made me feel.

One evening, after a long day at work, I sat on the edge of our bed, staring at my unopened wallet. I had money, yet I couldn't spend it freely. The thought made my chest tighten.

He noticed my silence. *"What's wrong?"* he asked, looking up from his phone.

I hesitated before speaking. *"Nothing,"* I said, forcing a smile.

But deep down, I felt pathetic. How had I gone from being independent to asking permission for everything? How had my life, once so free, become so restricted?

After a few months, once I had saved some money following our wedding in Egypt, we planned a trip to my home country. This would be the first time I introduced him to my family, and we decided to make it a surprise. No one knew we were coming.

It was the month of Ramadan, and we had planned to stay until Eid. With the help of my close friends, I secretly organised everything—our arrival, a family gathering, and a small wedding reception to celebrate with my loved ones.

When we finally arrived, the reaction was everything I had hoped for. My mother stood frozen in shock before her face lit up with joy. The house was suddenly filled with laughter and excitement, with everyone rushing to welcome us. But amidst all the warmth and familiarity, something still felt off. I thought being back with my family would allow me to be myself again, but I quickly realized that wasn't the case.

Three days after our arrival, it was time for the wedding reception. Everything was set—the food, the decorations, the guests were on their way. But just half an hour before the event, I found PIA still lounging around, not showered, let alone dressed.

I approached him carefully. *"You need to get ready. Everyone's waiting,"* I said, trying to calm my voice.

He looked up at me, his expression instantly darkening. His sharp, piercing stare sent a chill through me. He didn't like being told what to do.

I swallowed my frustration and softened my approach. *"Please, just for my parents, for my family. They've all come to celebrate with us,"* I pleaded.

He let out an exasperated sigh but finally got up and changed into a simple Arabic *kandura*—not even a wedding outfit, just something presentable. I felt relieved, hoping that was the end of it.

During dinner, he barely spoke. He ate silently, avoiding eye contact, and as soon as the meal was over, he disappeared into the bedroom without acknowledging anyone. My heart sank. My family and friends had gathered, full of warmth and joy, yet my husband couldn't even pretend to be part of it.

When the last guest left, I went to our room, exhausted but relieved that the evening had gone smoothly despite his behaviour. But the moment I stepped inside, his words shattered me.

"Don't sleep on the bed with me," he said coldly.

I stood there, stunned.

"This is my house. My room," I whispered, barely believing what I was hearing.

Without another word, he grabbed his things and slammed the door behind him. Moments later, I saw he had gone to sleep in the car from the window.

My father must have noticed. Sensing something was wrong, he asked me, "What happened?"

I forced a smile. *"He just needs the air conditioner,"* I lied. *"It's cooler outside."*

But my father wasn't convinced. His eyes, filled with quiet wisdom, studied me for a long moment before he turned and walked outside. I held my breath as I watched him approach the car.

Minutes later, PIA walked back inside. He obeyed my father's request and returned to the room. But once we were alone, he turned to me and said, *"Sleep on the floor."*

I felt my heart shatter into pieces.

Tears burned my eyes, but I refused to let them fall. I lay in the darkness, curled up on the cold floor, feeling more alone than ever.

Only God knew how broken I was at that moment. Thank God, our trip to my hometown lasted only two weeks—unlike Egypt, where we had stayed for an entire month. As much as I cherished the brief time with my family, I felt a strange sense of relief when we boarded the plane back to Dubai on the third day of Eid.

Returning to the city's vibrant yet unforgiving rhythm, I slipped back into the routine of hotel life. To outsiders, working in a luxury hotel seemed glamorous—polished uniforms, grand interiors, warm greetings exchanged in elegantly lit lobbies. But the reality was far from the illusion: long hours, relentless demands, and barely enough money to justify the exhaustion.

One afternoon, I received an unexpected call from Abu Dhabi. It was for a job interview. I had no idea what company it was or how much they offered. I only knew one thing—this was an opportunity, and I was willing to explore it.

I mentioned it casually to PIA over dinner. *"Let's just see what they offer. If the pay is good, we can consider it,"* I said, trying to gauge his reaction.

I should have known better.

A week of endless arguments, cold silences, and emotional manipulation followed.

"Why do you need to go?" he demanded, his voice sharp with suspicion.

"It's just an interview," I reasoned. *"I'm not even sure if I'll take the job."*

He scoffed, shaking his head. "You think I don't know what you're doing? You want an excuse to leave me behind."

His accusations stung, but I refused to engage in another pointless fight. Every conversation turned into a battle, and I was exhausted.

For days, we debated. He twisted my words, questioned my loyalty, and made me feel guilty for even considering a better future. It was mentally draining, but in the end, I stood my ground.

"We're just going to see what they offer," I said firmly. *"No decisions yet."*

Reluctantly, he agreed. But the unease lingered, settling like a heavy weight in my chest.

Deep down, I knew this wasn't just about a job. It was about control. And I was beginning to realise how much of myself I was losing.

The interview went better than I had imagined—so well, in fact, that it almost felt too good to be true. I got the job—a fresh start, a chance to build something better for myself. The salary was double what I was making, enough to finally leave the cramped staff accommodation and move into a place of our own.

But, of course, he wasn't happy.

PIA disagreed with my decision. He argued, dismissed my reasoning, and clarified that my independence unsettled him. But I had already made up my mind.

"This is for us," I told him, my voice steady, though inside, I felt exhausted from all the convincing. *"We can finally live like a real couple, not in staff housing."*

I knew he had no money. If I waited for him to provide for me, we would remain stuck in that tiny room forever. So, I took it upon myself. I used my savings for the house deposit and committed to covering the rent with my salary. He was supposed to provide shelter, food, and clothing as a husband—but in reality, he provided nothing.

The day finally came. I moved into a new studio apartment in Abu Dhabi. It was still small, but at least it had an attached bathroom and a tiny kitchen. It was mine, ours, a place to breathe.

Except there was nothing to fill it.

No furniture, bed, or dining table—just an old sofa that doubled as a mattress and my suitcases full of clothes. Life was starting over from zero.

For the first three months, I commuted back to Dubai every weekend. The journey was tiring—an hour and a half on a bus, then another transfer to reach his place. But I made the effort. I wanted to see him, to make things work.

Yet, the distance did nothing to soften him. If anything, it made him worse.

His temper grew shorter, his jealousy sharper, his ego more fragile. Small things triggered days of silence. I never knew what would set him off—one wrong word, one innocent conversation, one decision he didn't approve of.

And then, there was his birthday.

I wanted to surprise him. After work, I traveled back to Dubai, carrying a small cake and a gift. When I reached his accommodation, I carefully opened the door, expecting an empty room. I had planned to decorate it to make him feel special.

But the surprise was mine.

He was there. Not at work, not in Abu Dhabi visiting me—just sitting in his room, like he had nowhere else to be. I hesitated for a moment before stepping inside. His eyes met mine, cold and unwelcoming. Then, without a word, he stood up, grabbed his keys, and walked toward the door.

"Where are you going?" I asked softly, my hands still holding the decorations.

He didn't stop. Didn't answer.

Before leaving, he threw out one final remark. *"I don't want to see you here when I return."*

And just like that, he was gone.

I stood there, frozen, my mind struggling to process what had happened. My heart pounded, but my body refused to move.

I had come all this way.

For him.

I could have left. Walked away in anger. But instead, I swallowed the pain, wiped my tears, and quickly set up the decorations. I placed the cake and gift on the table, making sure everything looked perfect. Then, without another moment, I called for a taxi and left in silence.

That night, I ignored his calls. Maybe he had come back to find the room decorated, the effort I put in despite how he treated me. Maybe he felt guilty. Maybe not.

But I knew one thing for sure—I felt useless. Defeated.

I was his wife, not his slave. And yet, with every passing day, I was starting to feel like I didn't belong anywhere.

I don't remember a single holiday without a fight.

One particular trip stands out. It was our second year of marriage, and, as usual, we spent a month in Egypt. We had just left Sharm El-Sheikh, waiting on the roadside for a bus. The air was thick with the scent of dust and gasoline, the heat pressing against my skin like a heavy blanket.

I was conversing casually, trying to pass the time, when I mentioned that my phone battery was overheating. He barely acknowledged me, brushing off my words as if I were exaggerating. Wanting to prove my point, I gently pressed the warm device against his cheek.

Before I could react, his hand flew across my face, a sharp, stinging, shocking slap. The sound echoed in the street, louder than I expected, or maybe it just felt that way because I was too stunned to move.

Passersby stopped for a moment, their eyes flickering with curiosity before they hurried along, pretending not to see. My face burned—not just from the impact, but from humiliation. I stood there, frozen, as he turned and walked away without a second glance, leaving me alone on the roadside.

For a split second, I thought about running. I imagined myself rushing to the airport, buying the first ticket out. But reality hit just as fast as his hand had—my passport was with him. The sun was setting, and the unfamiliar streets around me darkened by the minute.

Twenty minutes passed before his younger brother arrived. He must have seen the fear in my face, the way my hands trembled slightly at my sides. He didn't ask what happened. He didn't need to. Without a word, he took me home.

At the house, I forced a smile. I spoke only when necessary, careful not to let my voice waver. I acted like nothing had happened as if my face didn't still tingle from his palm.

If he wasn't fighting with me, he was fighting with his family. Their loud and fast conversations often blurred the lines between joking and arguing. Arabic words flew across the room, sharp and heated, and though I didn't understand everything, I caught enough with the help of his siblings.

It became clear that this was their way of life—anger woven into their daily routine, arguments as common as meals. They fought like enemies, then laughed like friends, as if conflict was just another form of communication.

That night, as I lay in bed staring at the ceiling, I wondered if this was really what marriage was supposed to be.

For all those years, I believed it was my mistake. I told myself that I wasn't patient enough, that a good wife should endure, obey, and wait for her reward—either in this life or the next. I convinced myself that marriage had its ups and downs, and maybe, just maybe, the good moments would outweigh the bad.

There were moments of laughter, moments when we did silly things together. Every year, on my birthday, he had a routine. A month or a few weeks before the day, he would take me to a gold shop, choosing a gift for me—but never allowing me to pick it myself. It was always his choice, not mine. And the strangest part? Once we left the shop, he wouldn't hand me the gift. Instead, he would keep it, waiting until my birthday to present it as if it were a grand surprise.

Every year, the same performance. In the middle of the night, while I pretended to sleep, he would sneak out to his car, retrieving a balloon or a small bouquet. I knew exactly what he was doing, but I never let on. I would keep my eyes closed, listening to his exaggerated footsteps, waiting for the moment he'd gently place the gift beside me, whispering, *"Happy birthday."*

Six years of the same routine. Six years of predictable gestures. What once felt endearing soon became an empty ritual that lacked real thought or understanding of who I was. Yes, it was silly. But more than that, it was painfully empty.

One day, when he was in a good mood, I finally gathered the courage to express something that had been weighing on me for years.

"You've never given me nafakah," I said cautiously, choosing my words carefully. *"I know I earn my own money, but a husband has to provide for his* wife, even in the smallest way. Even if you gave me just ten dollars a month, I'd be happy."

To my surprise, he took my words literally. From that day on, he handed me exactly ten dollars each month—as if he were doing me a great favour.

We went grocery shopping together, but the division was clear even there. He paid for what he wanted, and I paid for what I could eat. Not what I wanted—because what I wanted was always *too expensive* or *unnecessary* in his eyes. He rejected my choices as if he were paying for them, even though that wasn't true. I felt like a guest, restricted in what I could buy despite working hard and earning well.

Even clothes required his approval. If I dared to purchase something without consulting him first, it would trigger a long, exhausting argument filled with harsh words. And if I stood my ground? Silence. Days—sometimes weeks—of silent treatment until I broke first. Not because I was wrong, but because I couldn't stand the tension. I couldn't bear the anxiety.

He decided everything. Who I could help. How much I could send to my parents.

I remember the day my father needed $5,000. Out of respect, I informed my husband—not to ask for his money, but simply to let him know. Instead of support, I got interrogation.

"Why do they need that much? Where is it going? Are they giving it to your brothers?"

In our culture, a daughter does not question her parents' needs. It was unthinkable for me to ask my father why he needed the money, yet my husband forced me to. I still remember the hurt in my father's voice when I asked him. I felt like a child begging for permission, not an independent woman making a decision about her own earnings.

After countless heated arguments, I sent the money anyway—but only $3,000, as my husband had *instructed*.

Looking back, I feel ashamed of how much control I allowed him to have over me. This pattern repeated every time my parents needed financial help. He treated it as if I were taking money from *his* pocket when, in reality, he had never contributed a single cent.

I was exhausted.

I had my own income, yet I lived with restrictions as if I had none. I worked hard, yet I had to fight for every little thing. And the worst part? I had accepted it for far too long.

Chapter 14:

"Carrying Hope, Bearing Pain"

By the end of our second year of marriage, the questions began.

"Why haven't you gotten pregnant yet?"

His words were not filled with concern but with quiet accusation. As if it were entirely my fault. As if my body had betrayed him.

At first, I tolerated it. I reminded myself to stay patient. I listened without reacting, knowing that his arrogance would never allow him to acknowledge the possibility that infertility could be a shared issue—or simply God's will.

When I suggested medical treatment, he refused outright.

"My mother never went to a hospital, and she had ten children," he scoffed. *"Why would we need doctors?"*

He spoke as if science had no role in life. As if God had left everything to chance.

Still, I didn't give up. I researched tirelessly, combing through articles and making lists of what could improve fertility naturally. When we went grocery shopping, I carefully selected nutritious foods that could help, but he always interfered. He picked through my choices, tossing out what he considered *too expensive* or *unnecessary*, allowing me to keep only what he thought was *acceptable*.

It was infuriating.

By the time I turned forty, my health began to decline. The pain became unbearable, and urgent care visits became routine. Doctors ran tests, but no one could tell me what was wrong. The only silver lining was my job at the hospital, where I had access to different specialists. I moved from one doctor to another, hoping for an answer.

After years of frustration and arguments, we finally agreed to try IVF. It wasn't cheap, but thanks to my staff discount, I managed to get more than 50% off. Still, even with the discount—*even with us splitting the cost*—the money became yet another argument.

At our first appointment with the fertility doctor, I finally received some clarity.

"Before we proceed, we need to perform surgery to clean your uterus," the doctor said, her tone gentle but firm. *"It's necessary before we can attempt pregnancy."*

I sat there, absorbing her words, feeling a mix of emotions: relief that we had found an issue, fear of what lay ahead, and, deep down, a quiet resentment because I already knew that, no matter what happened next, I would be facing it alone.

When the diagnosis came, it was already at an advanced stage.

It explained everything—the unbearable pain that left me unable to walk or stand every month, the repeated visits to the emergency room, the unexplained struggle to conceive. The fertility doctor referred me to an endometriosis specialist, one of the best in the hospital where I worked.

IVF wasn't just about the money. It was about trust. Hope. Fear. It was about enduring a process that forced me to confront my deepest vulnerabilities, all while being married to a man who saw no value in the journey.

Two weeks after my surgery, we returned to the IVF centre to begin treatment.

I will never forget the humiliation of it all—the endless tests, the invasive procedures, the sterile rooms where I felt more like a subject in an experiment than a woman trying to bring life into the world. The pain, both physical and emotional, became a constant companion.

The day of the embryo transfer was supposed to be hopeful. Instead, it turned into another fight.

"Why are we even doing this?" he snapped, his tone sharp with irritation. *"I don't even want a baby."*

I froze. My hands clenched into fists at my sides. This was the same man who, just weeks ago, had questioned why I hadn't gotten pregnant. The same man who had pressured me to bear him a child. Now, when we were on the verge of it, he was pulling away.

The contradiction was unbearable.

I had no choice but to continue. Every day, I injected myself in the stomach, pushing through the pain and nausea. With my history of endometriosis, the medication regimen was even more intense. I felt like my body wasn't mine anymore—it was just a vessel, subjected to treatments, needles, and hormones.

Then, two weeks later, the impossible happened.

The test came back positive. I was pregnant.

For a brief moment, everything felt worth it. The years of pain, the arguments, the exhaustion—none of it mattered because, finally, there was hope.

As part of the IVF protocol, I continued my check-ups at the fertility centre. Each visit filled me with cautious optimism. But at five weeks, something went wrong.

The bleeding started suddenly—heavy, relentless. Panic took over as I rushed to the emergency room, hoping, praying that there was something they could do.

There wasn't.

It was gone.

I sat there, numb, staring at the ceiling while the doctor's voice faded into the background. I didn't cry. I didn't scream. I just felt… empty.

Disappointed.

In denial.

It wasn't just about the money we had spent. It wasn't just about the physical toll. It was about losing the one thing I had held onto—hope.

Days passed, but the emptiness inside me remained. It wasn't just the loss of the baby—it was the loneliness, the weight of unspoken grief pressing down on me. I kept telling myself I needed time to heal, that eventually, the pain would dull. But the reality was cruel.

Instead of comfort, I was met with blame.

"This happened because of you," PIA said coldly. *"You shouldn't have been working. You shouldn't have been doing house chores."*

His words hit me harder than I expected. I wanted to fight back, to tell him that none of this was my fault. But I had no strength left.

He had never allowed me to share the pregnancy news with my family. And now, in my darkest moment, I wasn't allowed to grieve with them either.

I had nobody.

I thought about calling my mother, just to hear her voice, to feel some warmth in this cold reality. But what would I say? That I had been pregnant? That I had lost the baby? That I was hurting, alone, with no one to lean on?

Instead, I swallowed the pain. I went through the motions of daily life, pretending to be okay. At work, I smiled when needed. At home, I cooked, cleaned, and played my role. But inside, I was crumbling.

I didn't cry.

I couldn't.

Because if I let myself break, who would be there to help me pick up the pieces?

I wasn't living anymore—I was existing. A body moving through life, hollow and numb. A zombie trapped in a human body.

The problem with me was always the same. After every fight, after every harsh word, after every moment of emotional exhaustion, I was the one who offered peace first. Not because I was wrong, not because I wanted to, but because I was too tired to entertain his childish attitude. I was too sick of the endless drama, too drained from living in stress.

At some point, peace became more important to me than winning any argument.

So, I put my pain aside, swallowed my pride, and we tried again.

We still had 11 frozen embryos from our first IVF attempt. The doctor reassured us that they were of high quality and viable for up to ten years. This time, I wanted to give myself the best chance possible. I took a month off work, determined to rest properly, to avoid unnecessary stress, and to eliminate anything that could go wrong.

But IVF wasn't just about rest.

Every day, I injected myself with progesterone, along with other medications I could barely pronounce. My stomach turned into a battlefield, covered in deep purple bruises. My hip felt like it was on fire from the relentless shots. Some days, my skin was so numb that I couldn't even tell where the needle had pierced.

Yet, despite all this, I didn't have mood swings.

PIA did.

He started acting as if he was the one pregnant—complaining about nausea, claiming he had cravings. It was almost ironic. This was the same man who used to accuse me of pretending to be sick when I suffered through my menstrual pain.

"You're too weak," he would say. *"Other women can handle it. Why can't you?"*

Every time he dismissed my suffering, I silently raised my hands in prayer. *Ya Allah, let him feel what I feel. Let him understand just a fraction of this pain.*

God is Great. And now, he did.

Thirty days passed. My leave ended.

On my first day back at work, I felt something shift inside me—a sudden, heavy discharge.

Panic gripped me. My body froze.

I couldn't move, couldn't breathe, couldn't process what was happening.

My closest friend saw the fear in my eyes. *"I'll take you home,"* she said gently. But I shook my head. *"Don't leave me. Take me back to the hospital."*

I already knew.

And just like that, the thing I feared the most happened again.

At ten weeks, I lost the baby.

I was exhausted—physically, emotionally, spiritually.

There was no one to turn to, no shoulder to lean on. No family to comfort me. No friends to confide in. Grief sat heavy in my chest, suffocating me, yet I had to carry it alone.

For a day or two, PIA acted sorry. He spoke softer, his tone laced with something that almost resembled regret. But like a passing storm, it didn't last. Soon, he was back to his usual self—hurling accusations and throwing curses as if they were part of his daily sustenance.

His words cut deeper than my wounds, deeper than my loss.

My pain, both seen and unseen, became more relentless.

Before, the agony of endometriosis was confined to my menstrual days, but now it lingered—before, during, and after. There was no relief. No moment of ease.

Even intimacy became unbearable. What was once supposed to be a moment of closeness turned into sheer suffering. My body screamed in pain, but he didn't care. Instead, he punished me for it.

"You always have an excuse," he spat.

"You never think about what I need."

"You're useless."

More arguments. More shouting. More curses.

I began to resent him in ways I never thought possible. Sometimes, after he stormed out in the middle of a fight, I would find myself praying—not for peace, not for understanding, but for something darker.

"Ya Allah, if he is truly a curse in my life, remove him. Let him have an accident, let something take him away from me."

I knew it wasn't right. I knew it wasn't the kind of prayer I should be making. But in those moments, it was all I had. The only thing I could do was report everything I felt to God because no one else would listen. No one else would save me.

I carried the pain—both in my chest and in my body.

Even after the last surgery, nothing truly changed. Years passed, and the agony only grew worse. It was no longer just endometriosis; my fallopian tubes were blocked, and cysts kept forming, forcing me into an endless cycle of medical procedures. Every two years, I found myself back in the operating room. By the time I had my third surgery, I had learned to endure the pain without expectation, without hope.

And yet, we decided to try again.

We still had frozen embryos, a chance that felt both like a blessing and a curse. This time, we switched doctors within the same fertility clinic. We met with the head of the fertility department—an Egyptian doctor.

For the first time, PIA seemed engaged. Maybe it was because they shared the same mother tongue, culture, and beliefs. I let them talk, watching as they debated every possibility. I listened but stayed quiet. In the end, I followed whatever they thought was best.

This time, the process was even more rigorous. There were more tests, more medication, more injections, and more uncertainty.

But the doctor was different. He wasn't just focused on the science of IVF—he also understood its emotional weight. He taught us how to manage stress, handle emergencies, and prepare for what might go wrong.

For the first time, we felt something new.

Hope.

At six weeks, we heard it.

A heartbeat. Strong. Steady. Real.

For the first time, our hearts bloomed. The tiny sound filled the room, and for a moment, I forgot the years of pain, the losses, the emptiness.

Week after week, the baby grew. The next ultrasound showed the movements—tiny hands and feet kicking, a little body shifting inside me. It felt surreal.

But then, the doctor took longer than usual to complete the scan. His brows furrowed, and his focus deepened.

He turned to us and spoke carefully. *The baby's heart is positioned slightly lower than normal. The stomach, spleen, and colon are outside of the body, but this can change as the baby grows.*

My breath caught in my throat.

I wasn't sure how to process those words. Should I be relieved? Should I be scared? The doctor assured us that it was too early to panic. The organs might still shift into place. But we needed further monitoring.

It was one day before Eid, and we were supposed to travel home to celebrate. Instead, we rushed to book an appointment with a fetal medicine specialist at the hospital where I worked.

The next day, another ultrasound. Another doctor. Another explanation.

"Yes, the body hasn't fully formed, but this can still change." His voice was calm but firm. *"I need you to rest more. You're taking a lot of medication—why so many?"*

I hesitated before answering.

"I'm just following instructions."

But was I? Was I blindly trusting the process without questioning what was being done to my body?

Chapter 15:

"Ties That Test and Teach"

The night air in Abu Dhabi carried a familiar warmth, thick with the scent of jet fuel and the quiet hum of late-night travellers. I stood at the check-in counter, shifting my weight from one foot to the other, trying to ease the dull ache spreading through my lower back. My body was already exhausted, and the journey hadn't even begun.

PIA was in a heated discussion with the airline staff, his voice growing sharper with each passing minute. Our luggage was two kilos overweight—a small inconvenience, but one he refused to pay for. The staff remained firm, offering no exceptions, and I could feel the tension in the air tightening like an invisible rope.

"*We can just take something out,*" I suggested gently, trying to defuse the situation.

His head snapped toward me, eyes dark with irritation. "Stay out of this," he hissed under his breath, his face tightening in a way that sent a familiar chill down my spine.

I swallowed the lump in my throat and stepped back, watching helplessly as the argument dragged on. I wasn't even allowed to speak. I felt a dull sting—not just from his words but from the realization that, once again, I was invisible.

Unable to stand the tension any longer, I leaned toward another ground staff and whispered, *"I'm pregnant, and I'm not feeling well. Can you help us?"* I wasn't trying to cause trouble; I just wanted this moment to pass.

But when PIA realized what I had done, his expression darkened even further. He didn't want anyone to know I was pregnant. He believed that if we shared the news, people's jealousy would bring us harm. It was a superstition deeply rooted in his family, something passed down through generations. I didn't entirely blame him—faith and fear often blurred together in his world—but at that moment, I only felt exhaustion.

Finally, after an hour of arguing, the airline let us proceed without paying extra. By then, I was drained, my body screaming for rest.

As we approached the boarding gate, his mother's eyes met mine. She knew her son well—better than anyone ever could. One look at his clenched jaw and my tired, miserable face, and she understood everything.

"Don't argue with your wife when travelling. And never in front of people," she said firmly, her voice laced with quiet authority.

Under his mother's gaze, PIA's demeanour shifted. *"I'm sorry,"* he muttered, avoiding my eyes.

I nodded, not because his apology meant much, but because I didn't have the energy to fight. I forgave easily—it was a habit, a survival instinct. If I held onto every hurt, there would be no space left in me for anything else.

The atmosphere lightened as he turned his attention to his family. He enthusiastically explained our travel plans as if nothing had happened. I listened quietly, letting him take control. This was the start of our journey—a trip that had been planned for months. I didn't want to ruin it before it even began.

Their flight from Abu Dhabi to Cairo took three and a half hours, and then we boarded another plane to Kuala Lumpur—a gruelling seven-hour stretch. A three-hour layover awaited us before we finally reached our last destination.

It was Ramadan 2019, and we travelled at night, carrying a meal I had packed from home. It would be the last thing we ate before sawm (fasting) began. I sat by the plane window, looking out at the endless stretch of darkness, my hands resting gently over my stomach.

For a brief moment, I let myself hope. Maybe this time, things would be different.

We reached the final airport almost three hours before Iftar. The exhaustion from long hours of travel settled into my bones, but I reminded myself that this trip was supposed to bring joy. We had planned everything down to the smallest detail, ensuring a smooth journey—until we were stopped at customs.

A uniformed officer gestured for us to step aside after scanning our luggage. I glanced at PIA, who looked annoyed, his frustration evident in the way he clenched his jaw. His parents seemed confused. I took a deep breath and turned to him.

"Let me handle this," I whispered. "This is my country, and I speak the language."

His parents nodded in agreement. PIA, still sour from our earlier tension, crossed his arms but didn't argue.

With a practised smile, I approached the officer. "Is there a problem, sir?"

He eyed me before pointing at the screen displaying our luggage contents. "Are these your bags?"

I nodded. "They belong to my in-laws."

"Open them, please."

As soon as I unzipped the first bag, I almost burst out laughing. Inside, neatly packed in layers of plastic and paper, were bags of sugar, salt, flour, meat, and even vegetable fat. The officer frowned, clearly expecting something more suspicious. His serious expression only made it harder for me to hold back my amusement.

The absurdity of the situation made me chuckle. *Did they really think my country didn't have these basic groceries?*

Seeing my reaction, the officer raised an eyebrow. "What exactly is this?"

I turned to my mother-in-law, who was watching protectively as if the contents of her bags were precious treasures. "Mama," I said, shaking my head. "Did you pack an entire grocery store?"

She frowned. "You don't understand. The food here tastes different!"

I sighed. "Mama, every time I visit Egypt, I eat whatever is available. I never bring my own food."

"That's different!" she insisted. "We're used to our spices, our way of cooking."

The officer, clearly amused by the exchange, sighed and waved us through. "Next time, tell them your country has food," he muttered, shaking his head.

I nodded, smiling. *If only it were that simple.*

As we walked toward the exit, PIA remained silent. His mother, however, was still grumbling about how *no other food in the world tasted like Egyptian food.* I sighed inwardly. This was just the beginning of a trip that would test my patience in more ways than one. As we stepped out of the arrival hall, the rental car representative was already waiting, just as I had arranged the moment we landed. I was relieved—after such a long journey, I wanted everything to run smoothly. Without wasting any time, we transferred all our luggage into the car and hit the road.

The drive home was short, just twenty minutes, but it felt longer. The car was filled with quiet exhaustion. PIA drove the car, and I sat beside him, while his parents and sister sat in the back, murmuring in Arabic. I was wedged between them, the weight of responsibility pressing down on me.

Halfway home, we stopped at a convenience store to pick up toiletries. As I walked down the aisles, I caught PIA's reflection in the freezer glass. His expression was unreadable, his mood still unsettled from earlier. I let out a small sigh and grabbed what we needed, hoping that the rest of the trip would go more smoothly.

By the time we finally reached home, my family was already waiting. My heart warmed at the sight—this was a moment I had long envisioned, my family and his meeting for the first time. My mother stood at the doorway, her smile a mix of excitement and nervousness. My father, always composed, gave PIA's father a respectful nod. My siblings hovered nearby, eager but unsure how to break the ice.

The next few moments were filled with quick introductions, handshakes, and warm embraces. Laughter filled the air as we fumbled through translations—English, Arabic, and Bahasa weaving together in an unsteady but beautiful dance. My mother-in-law clutched my mother's hands, speaking rapidly in Arabic. My mother turned to me, eyes wide with amusement.

"What did she say?" she whispered.

I smiled. *"She's saying she's happy to finally meet you."*

My mother nodded, squeezing her hands in return. Despite the language barrier, their interaction was warm—a silent understanding between two women from different worlds, now bound by family.

After a few moments of chatter, I gently nudged everyone. "Alright, let's get you all settled in. You must be exhausted."

We drove to the rental apartment, where they could refresh before heading back to my family's house for iftar. As we entered, PIA's mother immediately began inspecting the place, murmuring in Arabic.

"What's wrong?" I asked PIA.

"She's worried about the kitchen," he said flatly.

I held back a sigh. *Of course.* The obsession with food never ended.

"Tell her not to worry. My mom has prepared everything."

He translated, and she nodded, though she still looked unconvinced. I chose to ignore it, focusing instead on the evening ahead.

As we left the apartment to return to my family's home for Iftar, I wondered if this was the beginning of two families truly coming together or just another test of patience and understanding.

Everything slowly fell into place, like the final pieces of a puzzle settling into harmony. The tension from earlier had faded, replaced by laughter, relief, and the comforting warmth of family. Even PIA seemed more at ease, his usual brooding expression softened as he clung to his parents and sister, speaking in rapid Arabic.

My parents' house was filled with life—my siblings, their spouses, nieces, and nephews. The chaos was overwhelming in the best way possible. Children ran around, their giggles echoing through the house, while the adults chatted over the clatter of plates and the fragrant aroma of home-cooked food.

As we gathered around the dining table, the exhaustion from travel was momentarily forgotten. There was no hesitation, no formality—just hands reaching for dishes, spoons clinking against plates, and the occasional burst of laughter as stories were exchanged in a mix of English, Bahasa, and Arabic.

I glanced at PIA's mother. She was eating quietly, occasionally glancing at the dishes my mother had prepared. I knew she was mentally comparing the flavours to what she was used to, but to my relief, she didn't say anything. Thankfully, the food she had insisted on bringing all the way from Egypt was safely stored—fresh meats and kofta in the fridge, dried goods tucked away in the rental house.

For the first time in a long while, everything felt... peaceful.

PIA was in his best mood, immersed in conversation with his father and sister, his usual edge dulled by the comfort of being with his family. I exhaled softly, allowing myself to relax, even if just for a moment.

After dinner, we excused ourselves and headed back to the rental house. The exhaustion from the long journey was catching up with us, and everyone was eager to rest.

As we drove back, PIA's hand rested on his knee, his fingers drumming lightly against his jeans.

"That went well," I said, breaking the silence.

He nodded. *"Yeah. My parents are happy."*

I smiled. *"That's all that matters, right?"*

He didn't respond; he just kept staring out the window. The silence stretched between us, comfortable yet heavy, as if both of us knew that this peace was fragile—temporary.

Still, for tonight, I allowed myself to believe in it.

The days passed quickly, each one bringing us closer to Eid. My pregnancy remained a secret, a delicate truth I carried within me while continuing my weekly IVF injections. Since I couldn't inject myself in certain areas, PIA and I had to sneak out to the clinic, where a doctor would administer the extra shots in my lower back. It was exhausting, but I endured it in silence.

Ramadan was nearing its end, and the house buzzed with last-minute preparations. The air was thick with the scent of freshly ground spices, sizzling meat, and the comforting aroma of cookies baking in the oven. Everyone had tasks—shopping for traditional clothes, cleaning, cooking—while I spent my time in the kitchen, my hands dusted with flour. Outside, children's laughter echoed through the neighbourhood, their excited chatter a reminder of the upcoming celebration.

For a moment, I allowed myself to breathe, wiping my hands on my apron and glancing at the golden trays of cookies cooling on the counter. The sun hung low in the sky, casting a warm glow over everything. It should have been a peaceful afternoon. Then the phone rang.

My brother's name flashed on the screen.

"Sis, are you home?" His voice was urgent, strained.

I wiped my hands again, suddenly uneasy. *"Yes, why?"*

"Check on our parents. They had an accident near the village entrance."

The words sent a chill down my spine. My heart pounded so hard I thought I might faint. I turned to PIA, and my breath caught in my throat.

Before I could even grab my headscarf, PIA and his father had already sprinted to the car. I ran after them, jumped into the backseat, and directed them toward the accident site.

We reached the scene in less than ten minutes, but it felt like an eternity. The road was crowded with onlookers, their faces painted with concern. The scent of burning rubber and gasoline clung to the air, mingling with the dust stirred up by passing vehicles.

Then I saw it.

My parents' car had veered off the road, its front crushed against a palm tree. The windshield was cracked, spiderwebbing out from the impact. My father lay on the ground, his face pale but conscious. My mother sat beside him, clutching her mouth. My sister crouched nearby, holding her young daughter, who sobbed quietly into her shoulder.

A man stepped forward, motioning for us to stop.

"Reverse your car. The road is blocked," he said firmly.

"That's my father," I replied, my voice shaking.

Immediately, others rushed forward to assist. We pulled over, and I ran to my father's side.

"Baba, can you hear me?"

He looked at me, his eyes alert but tired. My mother, on the other hand, kept her hands firmly over her mouth. When she finally lowered them, my stomach clenched—her front teeth were broken, blood smeared across her lips.

"We called an ambulance," someone informed me.

I glanced around, but there was no sign of one. My chest tightened. We couldn't just wait.

PIA acted quickly, clearing out the backseat to make space. With the help of bystanders, we carefully lifted my father, but just as we were about to move him, sirens wailed in the distance.

A medical officer jumped out of the ambulance, raising a hand. "Stop! We need to check for internal injuries before moving him."

We obeyed, stepping back as the paramedics assessed my father and lifted him onto a stretcher. My mother followed closely behind. Meanwhile, I hurried my sister and her daughter into our car, determined to follow the ambulance. PIA sped after them, the tyres screeching as he pushed the speedometer past 180 km/h. My pulse raced along with the car, but no matter how fast we went, we couldn't keep up.

By the time we reached the hospital, my neighbour was already there. His office was next door, and my brother had called him for help. I barely registered his presence as we rushed inside. The waiting area was sterile, the walls painted in dull shades of beige. The air smelled of antiseptic, the sharp scent stinging my nostrils.

Hours stretched endlessly. My anxiety gnawed at me, my patience wearing thin.

"Can someone check on my sister and her daughter?" I demanded at the reception.

The nurse barely glanced at me, her expression indifferent. "We're handling it."

Handling it? I clenched my fists, my frustration boiling over. "They need medical attention now."

She sighed, clearly annoyed, but after some back and forth, a doctor was finally assigned. Slowly, my siblings arrived one by one. My brother, who worked as a medical officer, used his connections to speed things up.

Finally, I was allowed into the Yellow Zone.

When I stepped inside, my father was awake. His first words weren't about his injuries or the pain.

"Did anyone check the car?" he asked. *"The chicken, meat, and vegetables—did they take them home?"*

I blinked, momentarily stunned. *"Baba, you're worried about groceries?"*

"Of course!" he huffed. *"I bought everything fresh for Eid."*

I couldn't help it—I burst out laughing. The image of a fresh chicken flying from the trunk and hitting his head during the crash played in my mind just like how he said. He smiled weakly, knowing exactly why I was laughing.

Meanwhile, my mother had lost several teeth from hitting the dashboard. That's when I realized—neither of them had been wearing seatbelts. A lump formed in my throat. It could have been so much worse.

As the evening wore on, word spread. Neighbours, extended family—people who hadn't planned on coming home for Eid—showed up. A bad situation had somehow turned into an unexpected family reunion.

By nightfall, my father was discharged, and we finally headed home. None of us had eaten all day, and only when we arrived did we realize we had missed breaking our fast.

To my surprise, my mother-in-law had taken charge of preparing iftar for the entire family. Despite not knowing anyone and not speaking a word of English or Bahasa, she had stepped in when it mattered most.

I stood in the doorway, watching her move through the kitchen with quiet determination. Something inside me softened.

In the darkest moments, there are always small blessings—acts of kindness that remind us what truly matters.

That night, as I lay in bed, exhaustion settling into my bones, I thought about how fragile life was. How everything could change in an instant. My parents had walked away from that crash with injuries, but it could have been so much worse.

I turned to PIA, who was already half-asleep beside me. His presence was a steady comfort, a reminder that through every storm, I wasn't alone.

I whispered a silent prayer of gratitude.

And for the first time in a long while, I felt truly at peace.

The day we had all been waiting for finally arrived. For the first time, all my siblings were together, accompanied by their spouses and children. My parents-in-law had also come, along with my sister- and brother-in-law. It was meant to be a joyous family reunion, but fate had other plans.

Just as my brother-in-law arrived, chaos struck—my parents were in an accident. The excitement of welcoming him properly was overshadowed by the rush to handle the situation. We quickly explained everything to him, and being young, he simply followed along with whatever we planned. His quiet adaptability was a relief amid the unexpected turn of events.

Despite the rocky start, everyone gathered as planned. We were dressed in traditional Malaysian attire, except for my parents-in-law. They weren't comfortable wearing foreign clothes, and I respected that. I didn't want them to feel forced to fit into our customs—comfort mattered more than appearances. It was a culture shock for them to be celebrating Eid in a completely different way, far from the traditions they had always known.

In Egypt, Eid was simple: men and women, elders and children, all attended morning prayers in a public space, and once they returned home, the day became quiet. There was no special breakfast, cookies, or house visits—just a special lunch shared within the immediate family. There were no guests. There were no extra celebrations.

For us, Eid was the complete opposite.

At dawn, the men left for the mosque, while the women stayed behind, preparing a grand breakfast. The house was filled with the aroma of freshly baked cookies, spiced curries, and fragrant rice dishes. In my family, once the men returned, we had a special ritual before eating—a 15 to 20-minute "Marhaban", a tradition of praising God and the beloved Prophet in unison.

The rhythmic chanting, the heartfelt supplications, and the warmth of the family coming together always gave me a sense of deep peace and belonging.

Then came breakfast—a feast.

Our family was so large that a dining table simply wouldn't do. I counted 56 people that day, all seated on carpets and sheets spread across the large living room. The room buzzed with laughter and conversations as we passed plates of food, tasting everything. For my in-laws, this was overwhelming, yet fascinating.

I found myself explaining each dish:

"This is "Nasi Dagang"—steamed rice with coconut milk and fenugreek seeds, eaten with fish curry."

"This is "Rendang"—a slow-cooked beef dish with coconut and aromatic spices."

"And these cookies? We call them "Kuih Semperit". Try one!"

My father-in-law had made a special request long before our visit—he wanted shrimp every day, cooked differently each time. My mother-in-law, on the other hand, had a deep love for fish and wanted to experience different cuisines—Malay, Chinese, and Indian styles.

Ironically, my brother- and sister-in-law were allergic to seafood, so for them, it was all about chicken and beef. They were astonished by the quality of the food, comparing it to a five-star restaurant.

"I didn't expect home-cooked food to taste this incredible," my sister-in-law said, eyes wide in delight.

Hearing this, my mother beamed with pride. It was a small but meaningful validation of the effort she had put into cooking.

After the meal, we moved on to the next cherished tradition—seeking forgiveness from the elders.

One by one, we approached them, gently taking their hands in ours and pressing them to our foreheads as a sign of respect. Apologies were whispered, blessings were given, and warm embraces followed. Tears were shed—tears of love, reflection, and gratitude.

Then came the excitement of "Eidiya"—money gifts for children and teenagers. Their faces lit up with joy as they received small envelopes, giggling as they compared amounts.

Before the arrival of guests and extended family members, we managed to squeeze in a quick family photo—a rare and precious moment that captured not just a gathering but the fusion of two cultures, two traditions, and two families bound by love.

In that moment, I realized something.

Eid wasn't just about customs or rituals. It was about togetherness, welcoming differences, and creating shared memories despite our unique backgrounds.

And that, I thought, was the true beauty of family.

The second day of Eid is always a spectacle—a grand reunion where tradition meets indulgence. This year was no different. With my Egyptian family eager to experience another layer of culture, we set out on a journey to visit all the eldest relatives, starting with my last living grandmother's sister.

It wasn't just a visit—it was an event. We convoyed in a procession of cars, moving from one house to another like a moving festival. At the first stop, we were seven cars deep; by the second house, more had joined. At one point, sixteen cars filled the narrow street, a sea of well-dressed relatives spilling into every driveway.

Each house greeted us with an array of fragrant cookies, warm homemade dishes, and endless cups of tea, the air thick with nostalgia and the scent of fresh spices. Conversations were kept light—updates on life were exchanged in short bursts before we moved to the next home. It was a ritual of reconnection, a celebration of heritage and belonging.

By the time we finished the fourth house, exhaustion clung to us like the lingering scent of cardamom and rose water. Our bellies were stuffed, our feet ached, and as we collapsed onto the sofas at home, the weight of the day pressed us into a deep, dreamless sleep.

After two days of much-needed rest, we resumed our planned itinerary for my in-laws. The road trip north awaited us—a long journey mapped out with stops for sightseeing, food, and of course, prayers.

That's when the real drama began.

My father-in-law, a deeply religious man, believed in taking his time for prayers. He didn't just pray—he immersed himself in the experience, often lingering in the mosque, reading the Quran, admiring the architecture, letting time slip away as if the journey itself was secondary to worship.

This did not sit well with PIA.

The first time we stopped for prayer, PIA's patience cracked. *"Baba, can we just pray quickly and move on?"* he asked, irritation laced in his voice.

His father, unbothered, responded calmly. *"You can pray quickly if you wish. I will take my time."*

That was all it took. From that moment, the arguments became a soundtrack to our journey—in the car, at roadside stops, over meals.

"You don't respect other people's time!" PIA snapped one afternoon as we waited outside yet another mosque.

"And you don't respect prayer," his father countered.

Their voices rose, their tempers flared, and yet, nothing changed. Each time we stopped, his father took his time, and PIA raged in frustration. The cycle repeated at every single prayer stop.

I tried to meditate, soothe, and distract myself, but the tension in the car grew thicker with every kilometre.

Beyond the endless prayer debates, the trip had other surprises in store.

One afternoon, after finishing our prayers, we realized my father-in-law was missing. Panic surged through me as we searched the surrounding area. After nearly an hour, we finally spotted him—at the top of a mountain, exploring a Hindu temple.

"How did he even get up there so fast?" I muttered, half in disbelief, half in frustration.

Another day, my mother-in-law fell ill, forcing us to leave her behind in the apartment while we continued our planned visits. That, of course, led to yet another family disagreement.

Then came the fight between my teenage sister-in-law and my mother-in-law over a ticket to the Twin Towers' sky bridge.

"Why do we have to pay for this?" my mother-in-law questioned, disapproval etched on her face.

"For the experience!" my sister-in-law insisted.

"But it's expensive."

"For you, maybe. Not for us!"

Tension crackled between them, an argument over a simple attraction turning into a battle of principles and perspectives.

The final straw came when PIA fought with his brother over something as trivial as a lack of gratitude.

"You didn't even say thank you for this trip," PIA fumed.

His brother rolled his eyes. *"Do I need to say it out loud?"*

That fight lasted until the very end of the trip. By the time they all boarded the plane back to Egypt, exhaustion wasn't just physical—it was emotional.

Through all the arguments, the stress, and the chaos, I needed a moment of joy. So, for PIA's birthday, I quietly arranged a surprise.

As we drove to our Airbnb in Kuala Lumpur, he had no idea that I had been secretly coordinating with our host—a cake, small decorations, a little something to break the tension that had followed us like a shadow.

When we walked in and the lights flicked on, revealing the small yet intimate setup, PIA's surprise and delight made it all worth it. For that night, at least, the family was united.

No fights. No tension. Just a shared moment of joy.

the STORY *of* BETWEEN BLOOD AND DESTINY

Life is an unpredictable journey—a tapestry woven with love, loss, resilience, and transformation. This book is a testament to that journey, reflecting on the struggles, triumphs, and lessons that shape a person over time.

"Bound by tradition yet longing for freedom, torn between love and survival, this is the story of a woman who has walked through fire and emerged stronger. This book spans decades and continents and follows her life's journey through love, heartbreak, betrayal, and self-discovery.

From the innocence of young love to the depths of betrayal, from the weight of cultural expectations to the pursuit of personal happiness, the narrative unfolds deeply, emotionally, and compellingly. It explores the complexities of relationships—marriages that tested her spirit, friendships that shaped her, and unexpected connections that brought joy and sorrow.

But beyond the pain and the struggles, this is a story of resilience. It is about a woman who refused to be defined by her circumstances and who fought for her identity, her dreams, and her right to choose her own destiny.

Told with honesty, vulnerability, and an unwavering belief in fate, this book is an intimate reflection on life's unpredictable twists and turns. It reminds us that no matter how dark the road may seem, there is always a path forward—one shaped by destiny but ultimately one that we have the power to walk with courage.

Queen of Flowers

Queen of Flowers is a resilient soul whose journey spans cultures, heartbreaks, and spiritual growth. With a voice both tender and powerful, she shares her deeply personal story to inspire, heal, and remind readers that love, faith, and strength can bloom even through life's most testing storms.

It wasn't a perfect trip. Far from it. But as I lay down that night, hands resting on my pregnant belly, I reminded myself—even in the middle of chaos, there are moments worth holding onto.

The joy of Eid barely lingered beyond a day.

By dinnertime, while I was lying in bed trying to soothe my aching back, I heard raised voices echoing from the living room. It wasn't the usual playful banter—it had escalated. Again.

This time, the argument was over dinner. Everyone wanted to order out and enjoy a birthday meal at a nearby restaurant. Everyone, except Mama.

"I'll cook pasta for PIA's birthday," she insisted firmly. "These restaurant prices are ridiculous! I'll get a stomach ache if I eat outside food."

The irony wasn't lost on me—we were on vacation, supposedly celebrating, and here we were, bickering over a $10 meal. Her resistance turned into a dramatic guilt trip. She paced the hallway, rubbing her stomach and sighing theatrically as if it might explode from one bite of restaurant pasta.

They eventually dragged themselves to the mall, exhausted from the long drive, only to stand around complaining that there was nothing suitable on the menu. The tension was so thick it could have been sliced and served at the table.

I slipped away during the chaos, pretending I needed the washroom. But really, I was heading to the gift shop. I wanted to surprise PIA with something special. He didn't ask for anything, but I needed to show him love amidst all the noise. No matter how anyone treated me—cold, unkind, or dismissive—I chose to stay true to my values. If I began returning cruelty with cruelty, what would separate me from those who hurt me?

That night, after we'd all settled back at the Airbnb, I handed PIA the gift—wrapped simply but with care.

His eyes softened. "You shouldn't have... How much did you spend?"

"It won't break my bank," I smiled. "Let's just talk."

We sat in the dimly lit corner of the living room. For once, just the two of us. We talked about our relationship, our behaviour, our hopes. There was a tenderness in his voice that I hadn't heard in a long time. He thanked me—not just for the gift, but for standing by him through the noise.

The next few days, we resumed our journey, exploring new cities. But the tension never really left. PIA and his younger brother had stopped speaking entirely. A week had passed without a word exchanged.

As our vacation wound down, we returned to my parents' home. I needed one more IVF shot—on my bruised hip—and I craved the comfort of my mother's voice and my father's calm.

It was during this time that my mother's side organized a fourth-generation family reunion. Our home was chosen to host. We threw ourselves into preparations—three days of cleaning, hanging fairy lights, setting up tents, and organizing the catering. I took charge of the guest list, registration table, and photography. I was exhausted but fulfilled.

Family poured in from all over—cousins I hadn't seen in years, aunties who smelled like jasmine oil, toddlers clutching sugary drinks. The air buzzed with stories and laughter.

My family, with its kaleidoscope of heritage—Malaysian, Chinese, Vietnamese, Indonesian—stood proudly united in our faith and traditions. Most of us had almond-shaped eyes and fair skin, a fact that surprised my father-in-law.

As we stood outside watching kids chase bubbles in the yard, I overheard him ask PIA in disbelief,

"Who are all these Chinese people? They celebrate Eid too?"

"That's all her family," PIA *replied, chuckling.*

"Really? Are they Muslim?"

"Yes, every one of them," PIA *said proudly.*

"Her family is mixed—Chinese, Indian, even German—but they're all Muslim. This is her world."

It was a rare moment of clarity for him—an opening into understanding the world he had married into. I watched him take it in quietly, and for once, he said nothing more.

Chapter 16:

"Empty Arms, Full Heart"

Some journeys don't end the way we dream them to, but they still shape us in ways we never imagined. This chapter tells the story of my quiet battles, silent prayers, and the moments that broke me and built me back. It is a testimony to hope, loss, faith, and the strength I never thought I had. In these pages, I share not just what happened but also what I felt, learned, and carried forward.

Three weeks had flown by—filled with joy, tension, long drives, and brief moments of peace. When it was finally time to return, we all boarded the same flight back. At Abu Dhabi airport, we parted ways. My in-laws continued their journey home, while we exited into the warm, familiar air of the UAE. As soon as we stepped outside and they disappeared from sight, I finally exhaled. I was drained—physically, emotionally, and mentally—but I didn't have the luxury to rest for long.

The very next day, we went back to the fetal specialist. It was the follow-up appointment we had scheduled before the vacation—to check if the baby's internal organs had moved into the right position and if the overall development had improved.

I remember the air in the hospital waiting room was cold. Too cold. My fingers trembled not just from the chill, but from nerves. My heart pounded beneath my ribcage. This was our third IVF. We had done everything right—extra genetic testing, stronger medications, and round-the-clock monitoring. I held on tightly to hope. I *had* to.

The Fetal Medicine Unit looked futuristic, with its advanced 3D imaging machines and dimmed clinical lighting. The doctor, an elderly man with calm eyes and a gentle manner, began the scan in complete silence. The screen showed our baby in crisp detail—every tiny movement, every shadow of bone, and every heartbeat.

Twenty-five minutes passed. Each second felt like a slow drip of fear.

Then he spoke.

He explained everything with precision—the measurements, the typical fetal development milestones for Week 18, and how our baby's condition compared to what was expected. He spoke slowly and clearly, but the words became harder and harder to digest.

Our baby was diagnosed with **Cantrell's Pentalogy**—a rare and severe condition that included multiple defects in the diaphragm, abdominal wall, sternum, pericardium, and heart. An *omphalocele*, he called it, where abdominal organs protruded through the belly area. The baby's heart was not where it should be.

I stared at the screen. I wanted to hold onto joy, but it slipped quietly out of the room.

He must've noticed how hollow my face had gone. He gently placed his hand on mine and said, "You have tried everything. We've done everything. Let's leave it to the Almighty now."

He offered to refer us to a government hospital for a second opinion, where more specialists could provide their opinions. But he didn't sugarcoat the reality.

"If the baby makes it to delivery, it may not survive more than a year. And even if it does, there will be countless surgeries—none of which can be done at once. It will be painful for the child... and for you."

He paused and then softly added, "If you choose to let go, it must be before 20 weeks. You have two weeks to decide."

We left the clinic in silence. No tears. Not yet. Just the weight of an impossible decision pressing on my chest like a boulder.

Back at work, I asked a pediatric heart surgeon to review the scan and the report. His opinion mirrored that of the FMU doctor—medically, it would be an uphill battle with little chance of peace for the baby.

Still searching, still unsure, I turned to a friend—a neonatal nurse manager who had seen more life and death than most. She didn't give me a direct answer. Instead, she held my hand and said, "Pray a lot. Whatever decision you make, make sure it's with a peaceful heart."

Her words stayed with me.

Pray a lot.

So I did.

I prayed not just for strength but for clarity. I needed to hear from the One who created this life inside me because, at that moment, no human advice could carry the weight of what I had to decide.

The weight of the decision pressed heavily on us. PIA, seeking reassurance, turned to his superior—an old man he deeply respected, a figure almost as revered as his father. His response was blunt, cutting through any lingering doubts.

"You've already asked five medical professionals, and you're still asking me? Why? Are you hoping for a different answer? Do you think it's haram to end the pregnancy? It would be far worse to let a child be born into suffering when you already know the outcome. I've seen it happen in my own family—it's agony for the baby, and I assure you, it will be agony for you too."

There was no room for debate. The truth had been laid out before us time and time again. With heavy hearts, we followed the medical advice and proceeded with the referral to the government hospital.

The doctor at the government hospital performed another ultrasound. Unlike the advanced 3D scans we had seen before, this machine was older, grainy, and unclear. But what she lacked in technology, she made up for in experience.

"I've seen this before," she said softly, glancing at the screen. *"I've witnessed babies born with this condition. Ninety per cent don't make it past the first 24 hours. Even those who survive endure multiple surgeries, each one a battle their tiny bodies may not win. I can see you're strong, but this will break you. I don't want you to suffer more than you already have."*

Her voice was gentle, yet firm. Compassionate, yet unyielding. By then, I was already 19 weeks pregnant. The law allowed medical termination only up to 20 weeks. We had no time left. No more doctors to consult. No more tests to run.

On July 7, 2019, I was admitted to the government hospital. Alone.

PIA had just returned to work after our vacation. There were no more leaves to take, no one to hold my hand. I walked into that hospital by myself, my heart heavy, my body numb, my mind replaying every moment leading up to this.

A nurse led me to a quiet room. It felt cold and sterile, nothing like the warmth of the life I had been carrying inside me. After a few hours, the doctor placed a tiny tablet under my tongue.

At first, there was nothing. Just silence.

Then, the pain hit like a wave crashing against jagged rocks. A sharp, relentless agony unlike anything I had ever felt—not even my worst endometriosis episodes compared. My body convulsed, shivers rattling my bones. The bed beneath me trembled with my involuntary shaking.

I reached for the emergency button, my fingers barely able to press it through the pain. A nurse rushed in, her face filled with concern.

"Shh, shh... It's okay," she whispered, wrapping her arms around me as sobs tore through my body. *"You don't have to be strong right now."*

But I wasn't crying from the pain alone. I was crying because I was alone.

No husband. No family. No comforting hand to hold. Just me, a sterile hospital room, and the unbearable weight of a choice no mother should ever have to make.

The violent shivering began to subside. Whether the medicine had fully absorbed or my body was simply surrendering to exhaustion, I couldn't tell. I lay there in the sterile hospital room, my thoughts fogged by pain and medication. The loneliness clung to me like a heavy blanket, suffocating and inescapable.

By evening, PIA arrived. His presence should have been comforting, but I was too drained to react. Just as he stepped in, a new wave of pain seized me—this time different. It wasn't just cramping; it was deep, pulling, unbearable. Contractions.

I gasped, struggling to catch my breath.

"I can't breathe," I managed to whisper, gripping the sheets.

PIA rushed to call the nurse. She checked my oxygen levels. *"Everything is normal,"* she said. But it didn't feel normal. My chest felt tight like I was drowning in the air itself. Maybe it was the pain. Maybe it was fear. Maybe it was something else.

Hours passed in a blur of agony and exhaustion. The contractions grew sharper, stabbing at my insides like a cruel reminder of what was coming. Then, a shift—an unbearable pressure. I felt something descending.

I forced myself up and stumbled toward the bathroom, the only place where I could sit in a position that brought even the slightest relief. The nurses had already placed a specimen collection pan on the bidet.

I barely had the strength to call PIA. *"Get the nurse,"* I murmured.

Within seconds, three nurses rushed in. One crouched in front of me, her voice calm but urgent.

"Okay, I see the legs. Take a deep breath."

I inhaled shakily, my mind barely registering the words. One more contraction came, more intense than anything before. And then—relief. The pain stopped.

It was over.

They helped me back to bed as one of the nurses gently collected the tiny body, carefully cleaning and wrapping him before placing him in a baby's bassinet beside me.

I turned my head, my heart pounding but my mind still numb.

There he was. Tiny. Fragile. Silent.

His little mouth was open as if he had tried to take a breath. His eyes were closed. His fingers curled ever so slightly, his tiny body bent unnaturally at almost a 45-degree angle—just as the scans had shown. His organs, too, were just as they had predicted, exposed and vulnerable. But his nose—his nose was unmistakable. Just like PIA's.

I exhaled, staring at him. My voice came out unexpectedly, soft, detached. *"Maybe when I couldn't breathe earlier... that was when he died inside me. That's why his mouth is open."*

The nurses chuckled at my words. *"No, it doesn't work like that,"* one of them said gently.

We shared a quiet, fleeting moment of absurdity, a brief escape from the crushing weight of reality. But then, as PIA recited a small prayer over our son, the grief finally found me.

The sobs came without warning, shaking my body more violently than the contractions ever had.

The nurse gave us a few quiet moments before returning with soft, knowing eyes. Her voice was gentle, almost hesitant.

"Are you ready to hand him over?"

I nodded slowly, though my chest tightened as if my body was fighting the very idea of letting go. PIA stood beside me, silent but composed. He kissed my forehead before following the nurse down the hallway, his steps unusually heavy.

I watched the door close behind them, and suddenly, the room felt colder, quieter—and emptier. I curled to my side, holding my belly, which had only hours ago carried a life I never got to know. I didn't cry. I couldn't. It was as if my tears had dried up from the inside out.

PIA handled the rest on his own. The hospital van transported our baby to another government facility authorized to conduct the burial. There, he was washed, wrapped with care, and taken to a small cemetery about an hour outside the city.

No family. No friends. Just the hospital driver and PIA—burying a child no one had ever met but one we had dreamed of, prayed for and fought for.

When PIA returned to the hospital, it was already close to midnight. I was lying awake in bed, eyes fixed on the ceiling. He walked in quietly and sat beside me.

"I recorded a few moments," he said and offered me his phone.

On the screen, I saw a tiny white cloth, a simple grave, and prayers whispered into the desert wind. My throat tightened. A lump formed so solid that it ached. I pressed my hand against my chest as if that could somehow ease the pressure.

I didn't speak. I just handed the phone back and turned away, facing the wall.

The next morning, I was discharged. As I stepped out of the hospital gates, the sun was shining harshly—mocking the grief buried deep inside me. Around me, families were left with balloons, flowers, and newborns swaddled in pastel blankets.

But I walked out with nothing. No baby. No cradle waiting at home. Just emptiness.

There were no arms to hold me, no mother or sister to whisper, "You'll be okay," and no one to cook me soup or rub my back. Just me.

PIA went back to work the next day, as if life had simply resumed. As if we had just returned from a routine checkup.

But I—I was stuck in a fog. Denial wrapped itself around me like a dull blanket. I wasn't fully aware of what had happened. My body felt foreign like it still remembered being pregnant, but my arms had forgotten how to feel full.

I couldn't tell if I was grieving the baby, the dream, or just everything I had been through over the last decade. Probably all of it.

Sometimes, I'd sit in the quiet of our studio and listen for a sound I knew wouldn't come. A cry. A coo. Even silence felt too loud.

I was given fourteen days of sick leave, with an additional forty-five days of maternity leave. But none of it felt like rest. The silence around me grew heavier with each passing hour, and the walls of our apartment felt like they were closing in. There was no baby's cry to fill the air, no midnight feedings or soft coos. Only the deafening stillness of grief.

I broke down.

Sometimes it was in the shower, sometimes while lying on the bed staring at the ceiling. Other times it would come unexpectedly—just a whiff of the baby oil I had bought weeks ago or the folded onesie tucked in the drawer.

PIA started to notice. The light in my eyes had gone dim. I barely spoke, barely moved. One evening, he sat beside me, quiet for a while before finally asking, *"Would you feel better if you flew home for a bit?"*

My eyes met his for the first time that day.

"Yes," I whispered. *"Please."*

I immediately opened my phone and began searching for flights. The prices were high—last-minute bookings always were—but I didn't care. I told him I'd take it. I just needed to get out. I was afraid I might lose myself if I stayed any longer.

PIA agreed—not because he was entirely comfortable with the idea of me leaving so soon, but because a colleague had spoken to him privately. A father himself, he had gently advised PIA,

"Let her go. Let her be with her family. This isn't just physical pain—she's grieving something deeper than words."

The very next day, just three days after my loss, I boarded a plane alone.

I sat by the window, staring at the clouds as the aircraft cut through the sky. My body was sore, my heart even more so. I felt strangely empty but heavy at the same time. I didn't cry. Maybe I had no more tears left.

When I landed at Changi Airport, I didn't call anyone. I didn't post a status or let a single soul know I was on my way. From the airport, I dragged my suitcase silently to the taxi stand and gave the driver my home address. The city lights blurred as we sped past, each one flickering like pieces of a life that no longer fit.

When the taxi pulled up in front of our gate, it was late. I stood outside for a moment, taking a deep breath. I hadn't even thought about what to say. How do you start a conversation with *"I lost my child"*?

I knocked.

My mother opened the door, her eyes widening in shock.

"My daughter?" she gasped. *"What are you doing here? You were just home for Eid!"*

I froze. My suitcase still clutched in one hand, I looked into her eyes, trying to form words. But my lips trembled, and nothing came out.

She reached out and pulled me into her arms.

"My goodness... what happened?" she whispered, holding me tightly.

But I still couldn't speak. I didn't even know where to begin.

I sat quietly on the edge of my childhood bed, my suitcase still half-zipped on the floor. The familiar scent of home—lemongrass oil, fresh laundry, and my mother's cooking—wrapped around me like a soft blanket. But no warmth could truly reach the hollow inside me.

Later that evening, I finally told my mother about the loss.

Her eyes widened in disbelief.

"How far along were you?" she asked, her voice trembling with concern.

"How did this happen? Didn't you know you were pregnant? What about your period? Didn't you feel anything?"

Her questions came in waves, not out of judgment but worry. I gave her the simplest answers I could manage. I didn't tell her about the IVF journey. I didn't tell her this was the third time I'd lost a baby. Some truths felt too heavy to share.

Instead, I said, *"I've always had irregular periods, Ma. And my size... it never really showed. I didn't look any different."*

I looked away, not wanting her to see the truth in my eyes—that I had been silently suffering for a long time, living a life that didn't feel like mine.

She sat beside me, holding my hand. Her fingers were warm, strong, and familiar. *"Maybe it was all the stress,"* she sighed. *"You were just here, and now this..."*

She assumed the long travels, the emotional weight of dealing with my in-laws, had taken their toll. I let her believe that. I needed her care more than I needed her questions.

For the next few days, she took care of me like she had done for my sisters and sisters-in-law after childbirth. She cooked nourishing meals—fish soup with turmeric and ginger, warm rice porridge, and herbal tea. She made sure I ate on time, rested well, and didn't lift a finger.

One evening, she asked gently, *"How are you feeling now?"*

"*I just want to sleep,*" I murmured. It was the truth. Sleep felt like the only escape from the ache that never left my chest.

A week later, my aunt was called in to help with postnatal massage—a traditional Malay practice passed down through generations. She came with her herbal oils and soft hands, her voice calm and steady.

The massage was not just physical—it felt spiritual. She worked to release the trapped "old blood," to realign my womb, to bring my body back to balance. But more than that, she grounded me. The gentle kneading of muscles, the rhythmic pressure along my spine, the scent of coconut oil mixed with healing roots—it was all meant to help me feel whole again.

And I needed it.

Because what most people didn't know was that on the day I delivered, the placenta didn't come out with the baby. The female doctor had inserted her hand and tried to pull it manually. I screamed, curled up in pain. When it didn't work, she waited an hour. Still nothing.

Eventually, they rushed me into the operating theatre. It took three hours to remove it surgically. When I woke up, I was numb—not just from the anaesthesia, but from the realization that my womb was now empty, stitched up, and aching with silence only mothers understand.

So when my aunt pressed her palms gently against my belly and whispered prayers under her breath, I closed my eyes and let go—for just a moment—of the guilt, the pain, the questions.

For those three days, I let myself be taken care of. I let my body heal, though my heart wasn't ready yet.

After spending a few weeks at home, wrapped in my mother's quiet care and traditional healing, I returned to work. On the surface, I slipped back into routine. But deep inside, I wasn't the same woman. I was grieving quietly, rebuilding silently, still holding onto hope even when everything felt fragile.

Two months later, I gathered enough courage to try again. I wasn't sure if I was fully ready—emotionally or physically—but time was passing, and I still had a good number of high-quality embryos preserved in the lab. I told myself, *"Why not? I had come this far."*

The transfer went smoothly. My body responded well, and for a moment, I allowed myself to dream again. I kept my hopes low, yet somewhere inside, I hoped the Almighty might show mercy this time.

But when the result came back, it was negative.

No heartbeat. No second line. No miracle.

Even my doctor was surprised. *"Everything looked perfect,"* he said, scanning the file in disbelief.

"You responded well... the embryo quality, the timing... *I really expected this to work."*

I stared at the floor. That familiar sting in my chest returned.

He paused, then looked at me with the kind of compassion only an experienced doctor—and perhaps a father—could offer.

"In the world of IVF," he said gently, *"even with all the science, we are still in the dark about so many things. What do we know today? Maybe not even seventy per cent of the truth. The rest... we are still searching for answers."*

He folded his hands together. *"But as a Muslim, I must tell you this: we can try, we can do everything within our power. But never forget—life, death, and everything in between are in Almighty's hands. We are just passengers in His plans."*

His words didn't erase the pain, but they brought a kind of calm I didn't know I needed. I nodded slowly.

"I think you should rest," he said. *"When you're ready to try again, I'll be here. You're not alone in this."*

I left the clinic quietly that day, walking back to my car with the soft evening breeze brushing my cheeks. I didn't cry. I just breathed. In and out. In and out. Trying to believe that maybe, just maybe, this was not the end of the road—only a bend in it.

Chapter 17:

"The Day He Said 'I Divorce You'"

Since the last failed pregnancy, the topic has been buried in silence, like a wound no one dared to touch. We never brought it up again—not once. And just like that, we slipped back into our old rhythm—but this time, everything felt heavier. The silence between us was no longer peaceful; it was suffocating.

PIA returned to his usual ways—using every past good deed like a weapon in his belt. It was subtle at first, a few comments thrown in here and there. Then it became routine. His frustration built up like a dam, until one night, he let it burst.

"*You're good at keeping the house,*" he said coldly, not even looking at me.

"*Cooking, cleaning, managing things. I never had to ask. You've never asked for money or gifts... but when it comes to intimacy, you fail like a broken wife. Other men are satisfied. Why can't I be?*"

His words landed like cold water on my skin. I sat still, heart pounding, holding back tears and rage at once. I clenched my jaw to stop myself from saying what I really wanted to scream.

A maid? Really?

You say you can pay someone to do what I do? Then go ahead—but remember, I've done all of it for free. You give me $10 a month, and I never complain. A maid would cost you at least $2000, and she wouldn't come with loyalty, patience, or love. I was your wife—not your servant.

But I didn't say that. I just nodded quietly, not because I agreed—but because I was tired. After six years, I had learned the pattern. Nothing I said could make him understand. He didn't want understanding. He wanted power. So I gave him silence, my only form of resistance.

Inside, I was cracking.

He even started comparing me to his mother—how she handled things, how she behaved. At some point, I wondered if he had secretly lived another life behind my back—how else could he speak so confidently about what other wives do, or how other women serve their husbands? Where was he getting this script from?

His expectations turned into accusations. His words, once soft and promising, now cut deep and left invisible scars. Still, I didn't argue. Not because I didn't know my rights—but because peace had become more valuable to me than justice. I wanted quiet, even if it meant swallowing my pride.

The accusations didn't stop. Every day brought a new wave of suspicion, questions, and curses flung like knives over the phone. I tried to block it out—tried to remind myself that the world was in crisis, that I needed to stay grounded. But it all came crashing down on March 18, 2020.

His voice cut through the static of the line like a blade.

"You are no longer my wife. I divorce you."

I froze. For a moment, I couldn't even breathe. The world didn't slow down—it stopped entirely.

What?

My jaw dropped, but no words came out.

Was this happening over the phone?

And then—*click*. The call ended.

I stood there, staring at my screen. The silence around me felt heavier than any ICU room I had walked into. My hands trembled. I didn't cry. I didn't scream. I didn't even tell anyone. I just moved through the next few days like a ghost—headless, clueless, just existing.

There was no message. No apology. No explanation.

I remembered how I had sent him chocolates and flowers just a few days before when I heard he was unwell. It was a small gesture, a bit of kindness, something to say that *I still care, even if everything is breaking around us.*

But instead of gratitude, he lashed out again.

"Why did you send those things?" he shouted. *"Do you want people to think I'm sick? That I took money or gifts from a woman?"*

His words stung.

He was more worried about his *image* than the truth.

He was more concerned about what others might assume than the fact his wife still thought of him, even through the storm.

"If they find out I'm sick," he added, *"they'll think I have COVID. They'll let me go."*

I couldn't understand it. His logic felt twisted, childish even. But I was too tired to fight. Too exhausted to challenge his paranoia.

Suddenly, there he was—at the door again. His face was softer than I remembered, and his voice unusually low.

"*I'm sorry,*" he said. "*I was angry at someone else. I didn't mean it. Please... let's be husband and wife again.*"

I didn't know what to say. I had just begun to process the loss, only to be yanked back into the confusion. He admitted he acted out of anger that the stress got to him, and he lashed out at the one person who always tried to stand by him.

Still... I gave him a second chance.

We revalidated our marriage, following Sharia law. On paper, we were husband and wife again. But in my heart, something had shifted. Something was no longer the same.

And then, just when I thought I was drowning silently in my own grief, the world outside cracked open with its own chaos.

A strange virus started spreading across the globe. The news called it COVID-19. At first, it felt distant—a problem for faraway countries. But it didn't take long before borders closed, lockdowns began, and fear seeped into every corner of our lives. Suddenly, we were all prisoners—not just emotionally, but physically. The world went still.

Yet somehow, the stillness outside only amplified the noise inside our home.

The pandemic wasn't just breaking systems—it was breaking people. I watched relationships crumble under the weight of fear and frustration, and ours was no exception. I thought the crisis would bring us closer. Instead, it exposed the cracks we both tried to ignore for far too long.

At first, the lockdown felt almost like a forced vacation—something surreal happening somewhere far away. We stayed in, cooked more often, and watched the news like a TV series. PIA and I even had moments of closeness. The silence between us softened briefly, replaced by the anxious buzz of the unknown.

But calm doesn't last long in a storm.

PIA's mood shifted when rumours of redundancies at his workplace reached home. His anxiety came in waves—one moment quiet and withdrawn, the next irritable and blaming. The fragile peace we had built cracked under the weight of uncertainty.

Meanwhile, my world spun in the opposite direction.

Overnight, our hospital transformed. The scent of disinfectant clung to every surface. Tension settled in the air like smoke. We had new protocols, endless checklists, temperature checks at every entry point, and no clear answers. Fear lived in everyone's eyes—patients, staff, even the higher-ups who usually walked confidently.

I was suddenly at the reception, registering visitors, asking for health declarations, recording contact histories, and checking symptoms. The security team handled body temperature checks, but the heat of the UAE summer made everyone impatient. Tensions boiled over fast.

One afternoon, a large man—tall, loud, and intimidating—stormed up to the desk, demanding to see his wife. I checked the system. There was no such name. I asked calmly, "Are you sure this is the right hospital?"

He glared at me, shoved his phone in my face with a blurry photo, and then yelled—in Arabic—a string of curses ending in:

"Ya ibn el sharmoota."

The direct translation?

"You motherfucker."

I froze. Then, without a word, I turned around and walked behind the desk. I collapsed onto the floor, hands shaking. Tears streamed down my face as I buried my sobs behind the counter.

> I'd endured so much personally, but I wasn't prepared to be publicly humiliated by a stranger. Not here. Not when I was trying to help.

I wasn't the only one. I'd seen my colleagues—nurses, receptionists, even security guards—duck behind walls or into supply closets, wiping tears with shaking hands. This wasn't just a job anymore. This was war.

Some days, I watched elderly parents begging to see their children, who had just given birth. I longed to say yes. But I couldn't. Not when I knew what was at stake. Not when the world was mourning thousands in mass graves, and we were fighting to prevent the same.

One afternoon, a man approached the desk without a mask. I'd already spoken to five people that morning about the same thing, but I reminded myself to stay calm and polite. I gestured gently and asked, "Sir, *please put on your mask before entering?*"

He didn't speak. He just flicked his hand at me as if I were a fly.

"*Go away. Don't talk to me,*" he said.

My chest tightened. Heat rose to my cheeks—not just from the stuffy mask or the suffocating suit but from pure, simmering frustration—the disrespect, the entitlement. It stung worse because I was doing my job—to protect him, to protect others.

I turned to my colleague, my voice trembling, and said,

"*Watch that guy. Something feels off.*"

She nodded, and we alerted security.

Later, we found out who he was. He was a doctor—a surgeon from another clinic who operated in our facility—a man trusted to cut open bodies and heal people who couldn't even follow the most basic safety rules during a pandemic.

I was livid—not just for me but for everyone risking their lives in that building. I didn't expect anything to come of it, but I reported him to the COO—not for action but because I needed to say it out loud. I needed someone in leadership to hear that this wasn't okay, that people like me—front liners, invisible workers—deserved respect, too.

In time, visitor control became too chaotic, but I was transferred to assist the COVID-19 Task Force.

This was a whole new battlefield.

That night, I went home and sat in silence—again. No one asked how my day was, no one said, "I'm proud of you," or "Thank you for what you do." The world was clapping for healthcare heroes, but inside my home, I was invisible—again.

And yet, I kept showing up.

Not because I was intense—but because I had no choice. Because the only way not to fall apart was to keep moving.

Now, I coordinate swab testing efforts across the city—answering emails, directing calls, assigning clinics to companies, and preparing logistics with the corporate office and nursing team. I also help organize medical teams to travel across the country—to homes, labour camps, and even remote desert areas, hours away from the hospital.

The moment test results came in, I was responsible for calling people marked in red—positive cases. Each call was a storm of emotions.

> "But I live with my parents!"
> "I have a newborn at home—what do I do?"
> "How long before I die?"

Some nights, my phone wouldn't stop ringing. Even in the bathroom. Even while half-asleep. I had to respond—there was no one else.

> "Hello? Yes, this is from the hospital..."
> "Yes, you tested positive."
> "No, you're not going to die. We're here to help."

I was tired, but I didn't want to stop—not when people were crying for help, not when they needed a voice to guide them through the fear.

Our task force wasn't just doctors and nurses. It was all of us—florists, HR, housekeeping, call centre staff, security guards. And me, the *"queen of flowers,"* now arranging people instead of petals. Coordinating swab vans, not wedding bouquets.

Some days, we were short on staff, so I joined the mobile testing teams. We wore full PPE—layers of scrubs, gloves, goggles, and face shields—in 50°C heat. We didn't dare drink water, and there was no bathroom.

We began at 4 p.m. and swabbed people until 3 a.m. Sometimes the lines stretched for kilometres. We sent security door to door in apartment buildings, urging residents to come down. Our quota?

10,000 swabs a day.

We had seven vans.

That's over 1,400 people per team per day.

When we returned to the hospital, exhausted and drenched, we handed over the samples, sanitised everything, and headed to the changing rooms. Stripping out of our soaked scrubs felt like shedding another layer of ourselves.

My undergarments were wet with sweat, and my hair dripped. I'd never known exhaustion like this.

Mornings came too fast. I'd wake up unrested, throw on my scrubs, and head into another day where I had to be composed, alert, and strong. The world outside was breaking—and inside me, too, something was breaking. Yet in that sterile, high-stress environment of the COVID-19 centre, I found a strange kind of stability. There were clear rules. Protocols. A purpose.

Unlike my home, where silence stretched like an invisible wall between us, work gave me instructions and structure—something I could follow when everything else in my life had lost meaning.

PIA wasn't impressed.

He demanded I quit.

"You'll catch the virus."
"Stay home. It's not worth it."

We argued. I stood firm.

"If I resign, and you lose your job—how do we survive?"

"We're in a pandemic. No one is safe. Not even one inch of this earth is free from COVID."

He didn't understand that this wasn't just about a job. It was about duty. It was about people. I couldn't walk away—not yet. Not when I was needed.

He called me stubborn. I called it survival.

I was unravelling. There were nights I couldn't sleep—not because of the sirens or emergency calls—but because of the emotional numbness settling into my bones. Professionally, I was drowning in responsibility. Every beep from my phone meant another crisis, another patient, another decision that could mean life or death.

I didn't get to process grief, and I didn't have the space to feel anything. My compassion became mechanical, and my strength automatic. And yet, I kept showing up.

Looking back, I realise how much I lost during that time—not just a sense of safety or emotional support but also a part of myself that believed love could withstand anything.

Sometimes, love isn't about holding on. It's about learning when to let go—even when the world around you is already falling apart.

Just two weeks after staying home, PIA was asked to return to work. Once a symbol of luxury and calm, the hotel had now transformed into a government-assigned quarantine facility. Ironically, he became part of the pandemic response team—not on the frontlines, but tucked away safely in the IT department, spending his days between the quiet of his office and the confines of his accommodation.

My reality, however, was the complete opposite. The phone rang nonstop. I answered calls while in the bathroom, replied to emails half-asleep with my eyes barely open, and coordinated crisis after crisis on barely any rest. Life blurred into a loop of alarms, PPE suits, and emotionally draining conversations.

Then, Abu Dhabi closed its borders. Anyone travelling in or out needed proof of a negative COVID-19 result. For many, this was an inconvenience, but for us, it became a turning point in our marriage.

PIA refused the swab test. He didn't trust the system, the science, or anyone in charge. There was no vaccine yet, no solid cure, just fear and uncertainty. When he was denied entry into Abu Dhabi from Dubai, his frustration boiled over. He called me, furious that I didn't pick up on time—never mind that I was juggling three phones, overseeing crisis reports, and trying to calm families whose loved ones had just tested positive.

He didn't believe I was working.

PIA's trust issues resurfaced with a vengeance. *"You're always busy,"* he snapped over the phone. *"With who? Who are you with?"*

It hurt. Not just the accusation but the fact that even after everything, he still couldn't see how I was breaking inside, from stress, from loss, from exhaustion.

One night, he managed to cross the border—somehow. It was well past midnight when he walked through the door, his face shadowed with rage.

"Give me your phone," he demanded.

"What's going on?" I asked, already too drained to argue.

He scrolled through my personal messages, work emails, and even task force communications. His eyes scanned every name—Abood, Moe, Ahmed—asking why I spoke to men after hours and why certain names kept showing up.

"*These are my superiors,*" I said calmly. "*They lead the pandemic response unit. I have to report everything to them.*"

He didn't believe me. "*You're deleting things. You're hiding something. One of them was here, wasn't he?*"

His accusations were sharp. Unreal. As if he had erased every image of me in full PPE, drenched in sweat, comforting patients with no families allowed near. He didn't care about the tears I cried after seeing someone pass alone in isolation. He only saw me as someone betraying him.

Without warning, he snatched my phone again. This time, he cloned it—linked my Google account to his, mirrored every email, photo, and message, and installed trackers and backup systems so he could monitor my movements remotely.

I should've felt violated. Angry. Broken. But I didn't have the energy to react. People were dying. I was barely functioning. Let him read everything. Let him accuse me. I had nothing to hide.

He could control my devices, but not my purpose, my focus, the urgency in every call I answered, or the prayers I whispered for patients I would never see again.

And maybe that's what changed me the most during that time: realizing that not everyone will understand your fight, even the one you married. But in the chaos of a pandemic, clarity came like a slow sunrise—I couldn't afford to lose myself just because someone else was lost in their fears.

The situation was getting worse.

Each morning started the same: my team and I huddled around a screen, watching the live updates tick upward. Case numbers were climbing worldwide. Nationwide. Then we checked our hospital figures—our home ground—and felt the weight settle deeper on our shoulders. There were days over a hundred staff tested positive. Each of them was sent into 14 days of quarantine. That meant fewer hands, longer hours, and no room to collapse.

We didn't talk about it much—but we saw it in each other's eyes. The weariness. The fear. The tight, forced smiles that said, *"I'm okay,"* even when we weren't.

Yet, it was during this time that I witnessed the strongest people I've ever met. Not because they lifted heavy equipment or worked sixteen-hour shifts without a break—though many did—but because, in spite of the chaos, they remained emotionally unbroken. We became each other's family. We took turns lifting each other up when someone crumbled. Sometimes, it was a hug; other times, it was just silent company while someone cried in a corner.

There were moments when we received calls—"Your father is in ICU," "Your mother has tested positive,"—and all we could do was whisper prayers across borders that remained firmly shut. That helplessness stays with you. It carves itself into your bones.

As a team leader, I didn't have the luxury of showing fear. My team needed strength—even if it was artificial. So I fed them hope, even when I was running on fumes. I smiled when I wanted to scream. I told them, *"We're doing great,"* when I didn't know if we'd survive the week.

The Health Authority's rules and protocols changed frequently—sometimes swabs were twice a week, sometimes only weekly. We learned to adapt and laugh when we could. I remember someone joked, "By December, our nostrils will be like tunnels from all the swabbing!" It was a silly comment, but we laughed. We needed that. We needed *something* to lighten the crushing silence between bad news.

Not everyone was kind. Some patients yelled at us, and others refused to wait their turn. One day, a doctor barged in again, demanding to be swabbed immediately. She waved her status like a badge.

"I have patients waiting—I don't have time to queue," she snapped.

I looked her in the eye and said calmly,

"Doctor, everyone here has patients. They're waiting too."

Her face flushed with anger. She stormed out and called my superior, complaining I was rude and shouting. I didn't raise my voice—I didn't bow.

She was known for her 'Karen' attitude, so no one took her complaint seriously. But the incident left a bad taste in my mouth. We were risking our lives and still had to fight for respect.

Things only got more complicated once the vaccine rolled in. There were more procedures, more questions we couldn't answer, and more frustration from the public. Every day was a battlefield—not of bullets but of exhaustion, miscommunication, and emotional breakdowns.

I was tired.

Tired of being strong.

Tired of smiling through fear.

Tired of being called rude for enforcing fairness.

Tired of being the one everyone looked to while I was quietly falling apart inside.

But I showed up. Every day. Because if I didn't, who would?

Chapter 18:

"Armor, Evidence, and Exit"

Each morning, I wore my uniform like armour, my mask like a second skin, and my smile like a shield. I showed up strong every single day.

To my team, I was a lighthouse in the storm—a leader who held it all together even when the numbers rose, when protocols shifted like quicksand and when the air around us felt heavy with unseen danger.

But beneath that armour, my chest ached. Not because I had tested positive for COVID, but because I was slowly breaking inside. The virus wasn't the only thing I was fighting. At home, I had become a ghost in my own marriage.

When we reunited after a long separation, I truly believed—hoped—that maybe things would change. For a brief moment, it felt like they had. Two months of fragile peace. But then it started again.

His words cut sharper than any scalpel.

"Work again? Who are you talking to for hours on that phone?"

It didn't matter that I was managing chaos in the hospital, leading a team through the worst health crisis of our generation. To him, my commitment looked like disloyalty.

Every time he got angry, he would throw the word around like a weapon.

"I'll divorce you," he'd say coldly. *"You're never around. What kind of wife are you?"*

The threats became a pattern—his go-to when things didn't go his way when I was too tired to talk, too busy to entertain his suspicion.

One night, after another baseless accusation, I said quietly, *"If that's what you want, then go ahead. Let's do it the proper way."*

He looked at me with a twisted smile.

"You go to the court," he said. *"But if you do, I'll make your life a disaster. I'll show you the real Egyptian."*

That moment? It didn't hurt anymore. Not in the way pain usually does. I didn't feel anger. I didn't feel fear. I didn't even feel hate. I felt... nothing.

I had become numb.

No love.

No hate.

No tears.

Just a silent scream inside my chest that no one could hear.

At work, I was the one people came to when they broke down. I was the listener, the motivator, the one who whispered, *"We'll get through this,"* even when I wasn't sure myself.

But who was I supposed to turn to?

One afternoon, my heart felt too heavy to carry alone. I told myself, *"Just find someone. Just talk."*

I headed to my superior's office, hoping for even five minutes of presence, of understanding. But her chair was empty. Instead, I turned to her right-hand man—the one everyone respected for his warmth and calm. He greeted me with the same cheerful energy he gave everyone else.

"Can I just sit here for a moment?" I asked, trying to hold my composure.

He paused, then studied my face carefully.

"Are you okay?"

I nodded. Then shook my head.

He quietly closed the door.

I didn't need to explain. He could see it in my eyes—the exhaustion, the cracks, the fight I was losing behind closed doors.

"I'm here," he said gently. *"If there's anything I can do..."*

And just like that, the dam almost burst.

I didn't cry. Not fully. But the tears burned the back of my eyes. I told him—everything. The fear. The threats. The loneliness. The emptiness of lying next to someone who made me feel invisible.

To my surprise, he didn't seem shocked. Instead, he looked almost... relieved.

"You know," he said, *"I've seen this in you for a long time. I could see it in your eyes—you've been carrying something you don't talk about."*

Then he added, *"I've been there too. I'm a divorcee. I know what this road looks like."*

He didn't lecture me. He didn't judge. He offered something more valuable: guidance.

Step by step.

Clear.

Organized.

Human.

What to expect.

What to prepare for.

How to stay grounded when the storm hits again.

And in that moment, for the first time in months, I didn't feel alone.

Before I ever confided in Abood, before I found the courage to admit I needed help, I had already begun my quiet search for a way out. Late at night, after my shifts ended and the house fell silent, I'd sit alone—my laptop casting a dull glow on my face—and scroll endlessly through government websites, trying to make sense of the tangled process of seeking help. Everything felt like a maze of Arabic forms, legal jargon, and impossible requirements.

I was tired of living in fear—not of the virus anymore, but of the man I came home to.

Abood didn't know it yet, but the day I walked into his office, I needed someone to acknowledge my pain. I didn't want pity. I didn't even want solutions at first. I just needed someone to see me—not the strong frontline leader, not the nurse who never broke, but the woman who was quietly breaking inside.

His guidance was simple but clear. He told me what steps to take, what to expect, and where I might find some clarity in the chaos. "Just one step at a time," he said, handing me hope in the form of structure.

So I started.

I documented everything—dates, words, threats. I filed applications online, wrote statements, and visited police stations. I sought advice from anyone willing to listen. The system, though, felt like a wall. For weeks, no response. Seven times I tried to contact the Judicial Department. Seven times, silence.

Then one morning, my phone rang.

I was called in for an appointment at the Judicial Department, just a kilometre from my home. I thought maybe, finally, something would move. Maybe they would understand.

But nothing prepared me for the way she looked at me—the woman who was supposed to help.

She sat behind a wide desk, her eyebrows slightly raised as she scanned my file. Her eyes flicked up at me, then narrowed. She didn't see a wife trying to save herself. She saw an older Asian woman as unimportant and unworthy of empathy.

Her voice was flat. "Have you ever been in a relationship before marriage?"

I blinked, unsure if I'd heard her right. "Yes," I said slowly, "I had a boyfriend. But nothing serious. Just a normal, respectful relationship."

"Did you have… a body relationship? Like husband and wife?"

The air in the room turned thick like something sour had been released. I sat straighter, my pulse rising. "I am Muslim. I know what is haram and halal," I said firmly.

She didn't flinch. "Sometimes men think their wives have someone else. Because of their history."

I clenched my hands under the desk, forcing my voice to stay calm. "Even if that's his belief, that doesn't give him the right to treat me this way."

She shrugged, not looking up.

"If you've already decided to divorce, then go through the court. We only counsel couples who want to continue."

I stared at her.

"I was referred here by the police after reporting verbal and emotional abuse. I didn't come for reconciliation—I came because I followed the system."

She closed the file.

"Then you need a lawyer."

Back to square one.

The lawyer I found was Egyptian. When I laid out everything—every form I had filled, every visit I made—he looked at me long and hard. Then he asked,

"Why did you marry this man if you have to pay for everything yourself?"

I didn't know how to answer that. Maybe love. Maybe foolishness. Maybe survival. Maybe hope.

He shook his head.

"What is this man thinking?"

he muttered, half to himself. Then he let out a long breath, opened the file I brought, and began marking it with a pen.

"This is what we will ask for," he said, listing out each item carefully:

"Mahar" – the dowry he never paid in full.

"Nafakat Edah" – the financial support I was owed during the waiting period.

"Muta'ah" – compensation for the emotional damage, for the years I spent enduring silence and threats.

"Khula" – divorce initiated by the wife, with compensation.

I nodded, taking it all in. There was no relief, not yet. But there was direction. And in that, I found something I hadn't felt in a long time—power.

I looked him in the eye and asked the question I'd been dreading.

"How much will it cost?"

The lawyer gave a short, almost apologetic laugh. *"We're expensive,"* he said with a shrug, like he'd said it too many times before.

"I understand," I replied.

"I'll figure something out. But please—be honest with me. I need an accurate number."

He leaned back in his chair, interlacing his fingers.

"Thirty thousand dirhams. That includes everything—from filing the case to getting the divorce papers. You won't need to lift a finger. We'll handle it all."

The amount stung, but I nodded. I wasn't shocked. I had been paying the emotional price for years—this was just the financial one.

"But," he added,

Try one more time. Apply online again. Add everything we discussed—list the dowry, the maintenance, the waiting period, everything. If there's still no response, come back to me. I'll help you. I want to fix the mess that man made—for your sake and for the dignity of our name.

I appreciated that—this unexpected loyalty from a stranger who happened to share my husband's nationality. So I did as he advised. I applied again, adding every detail, every injustice, every unpaid obligation.

Days passed. Then, one morning, as I sat in my usual corner of the apartment with a cup of black coffee, my phone buzzed.

An email.

From the Judicial Department.

A link for a video hearing.

I didn't celebrate. Not out loud. I'd learned to keep my hope quiet. But inside, I whispered *Thanks God*.

The first hearing was brief. He didn't attend. He didn't respond. The judge simply noted the complaint and adjourned.

The second meeting was different.

This time, his face appeared on the screen. The man I had once loved. The man who had once said I was his everything. He sat there, defiant, arms crossed, eyes darting like he was preparing for war.

The judge listened as I spoke. Then turned to him.

"Why have you not provided nafakah?" he asked.

> *Nafakah*—the basic financial maintenance a husband is religiously obligated to give his wife. It covers food, shelter, and clothing—essentials that Islam commands to be fulfilled with dignity. But for me? For years, I received nothing but crumbs.

He argued back, quoting verses and religious rulings as if he were a scholar. As if he knew Sharia better than the judge himself.

He rejected every claim. He denied everything. And so, the judge scheduled a third session.

By then, I was tired. My soul was threadbare. I wasn't chasing justice anymore—I was chasing peace.

At the third hearing, the judge got straight to the point.

"Do you both agree to divorce?"

There was a heavy pause. I stayed silent, waiting.

Then, from his side of the screen, he finally spoke.

"I will divorce her," he said, *"but only if she drops everything*—the *mahr* (dowry), the *nafakat iddah* (the support during her waiting period), the *mut'ah* (compensation for the harm caused), and the *khula* (the financial consideration for a divorce initiated by the wife)."

My heart didn't race. My hands didn't shake.

I simply nodded. *"Yes,"* I said calmly.

"I will drop it all."

Not because I was weak. Not because I didn't deserve every last dirham. But because I was tired. Because I had carried this burden for too long, and I was ready to put it down.

I added, *"For the seven and a half years we were married, he gave me only ten dollars a month for the first three years. Then thirty dollars a month. That was it. I have no expectations now."*

The judge looked at me, then at him.

"She has agreed. Now you must say the words."

He nodded.

And the judge began, *"With the God name..."*

With solemn instructions, he guided him through the declaration.

And just like that, with a few words spoken through a screen, I was divorced. Officially. For the second time. From the same man.

As the call ended, I didn't feel triumph. I didn't feel sorrow.

I felt... nothing.

No tears. No anger. Just a vast, echoing emptiness.

But as I stood up and walked across the room, something had shifted.

My steps were slow.

But lighter.

That final hearing happened on January 13th, 2021—just shy of a year since our initial separation.

But let me take you back a few weeks before that day, when the situation began to spiral into something darker, something far more humiliating than I was prepared for.

PIA had been acting erratic—paranoid, accusing, irrational. He was convinced I was having an affair. Not just with anyone, but with a colleague named Abood.

The accusation alone was absurd. Abood was not just a co-worker; he was someone I respected, someone who treated everyone with kindness and maintained firm professional boundaries. He was also a certified life coach—measured, composed, and emotionally intelligent. But in PIA's imagination, Abood had become the villain in a story he'd written in his own mind.

One morning, as I was sipping my second cup of bitter office coffee, my phone lit up.

It was PIA.

"I'm outside. I want to see Abood with you. Now."

I froze.

Not out of guilt—but disbelief. He had shown up. At my *workplace*. Uninvited. Unannounced. As if our private war belonged in the professional space I had worked so hard to protect.

I messaged Abood immediately: *"Please be ready. My ex is here. He's coming to see you."*

A few minutes later, we walked into Abood's office together.

PIA entered, chest puffed, eyes sharp with suspicion, walking like he owned the room. The air changed the moment he stepped in—thick with tension, like static before a thunderstorm. My heart pounded so hard I thought it would echo in the silence.

He looked at Abood with contempt and launched straight into his monologue.

"You're trying to destroy my marriage," he said, voice loud and laced with accusation. "You're the reason she's leaving me."

I stood there, burning with shame. Not because I had done anything wrong—but because this private wreckage was now scattered across my workplace floor, in front of someone I respected. My personal life had been dragged into the fluorescent-lit halls of professionalism, and there was no undoing it.

To his credit, Abood didn't react with anger. He listened and kept his tone calm and respectful.

"I'm a life coach," he said.

"I help people find clarity—not break marriages."

But PIA wasn't there to listen. He wasn't seeking the truth—he was building his case in his head, line by imagined line. No explanation would have satisfied him, because he wasn't looking for one. He had already decided who the villain was.

Later, he demanded my business phone. "I want to check everything," he said.

I refused. That phone didn't even belong to me—it was company property, used by the entire team. But reason didn't matter to him.

From that moment on, I knew I had to be extremely cautious. The divorce might have been finalized on paper, but in reality, I was still under surveillance.

I lived in a constant state of alert. Every day after work, I'd step out of the taxi and scan the street. I'd circle around the villa slowly, pretending to adjust my bag or check my phone, all just to make sure he wasn't lurking somewhere in that white sedan with the familiar plate number.

Then I'd unlock the door and quickly slip inside. Every time I heard keys jangling in the corridor, my entire body would stiffen. Even if I was asleep, the faintest metal clink could yank my soul out of my body. I would hold my breath, heart racing, waiting for a confrontation that, thankfully, never came—but always felt one second away.

It wasn't just the sound of keys.

It was the iPhone ringtone he used—the one that played like a soundtrack to my anxiety. It was the sight of a similar white sedan. Certain colognes were in the air. Even random WhatsApp notifications made my stomach twist.

These triggers lived rent-free in my nervous system for nearly two years. And the saddest part?

I never spoke about it. Not to a therapist. Not to a family member. Not even to the friend I carpooled with every day to work—the one who always asked how I was doing and meant it.

I wore my silence like armor. Because I didn't know how to explain what it felt like to live in fear of someone who once called me *his wife*.

Exactly one week after the court hearing, on my birthday, I walked into the office with a folded piece of paper tucked in my bag—the official divorce certificate. The ink was fresh, but the weight it carried was ancient.

I hadn't told anyone yet. Not even Abood. Not until today.

There was something symbolic about it—sharing the news on the same day I was born, as if I, too, was being born again. A quiet rebirth after years of emotional death.

I walked toward their office slowly, not even sure what I was feeling—numb, maybe. Relieved? Tired? But I knew one thing: I was ready to speak.

When I stepped into the room, Pei Pei was finally there.

For months, she had been just a name—someone I was supposed to meet on my first day, but her seat had been empty then. In her absence, I met

Abood instead—the man who had become an unexpected witness to my quiet unravelling and eventually a silent supporter.

Now here they were. Both of them.

I cleared my throat and held up the folded paper.

"I'd like you to read this," I said softly.

Pei Pei took it first, her eyes scanning the lines quickly. Abood leaned over to read with her.

And then, without warning, Pei Pei let out a high-pitched scream—not from shock, but pure, unfiltered joy. Her face lit up, and she jumped up from her chair.

"Oh my God, you did it!" she shouted.

She wrapped her arms around me before I could react. Abood grinned and joined in the hug. I stood there for a second, stunned, and then I allowed myself to melt into the warmth. It felt strange—being held like that. I celebrated. I believed.

We pulled apart, all three of us smiling with teary eyes.

"Happy birthday to you," Pei Pei said, placing both hands on my shoulders. *"This is your freedom day. We must celebrate. Dinner's on me tonight!"*

I nodded. Words felt unnecessary. My body was exhausted, but my soul felt a flicker of light I hadn't felt in years.

That moment—the hugs, the screaming, the joy in their faces—was more than just a celebration. It was recognition. It was a reminder that my pain was real and that surviving it mattered.

For so long, I had lived in silence. I had walked through each day wearing a mask, never letting the truth slip. But here, in this small office with two people who had seen bits and pieces of my storm, I finally felt safe enough to say, *It's over.*

I didn't know it then, but that moment would become a turning point.

A quiet promise that life could be different from here on.

That evening, Pei Pei insisted on taking us out for dinner. Her joy—and Abood's—was louder and brighter than even my own.

I watched them as we sat at the little Asian place tucked behind our office building, their faces glowing with pride and warmth. For a while, it was like watching two friends celebrate someone else's victory. But slowly, I allowed myself to believe—maybe this time, it *was* my victory too.

We ordered hot laksa, our breaths fogging up slightly as the steam rose from the bowls. The aroma of lemongrass and chilli filled the air, comforting and nostalgic. It reminded me of home, of simpler times, long before my name ever ended up in courtrooms or police logs.

Between slurps of broth and bites of noodles, the conversation began to flow freely. We laughed, paused, and reflected, breaking years of silence with stories that had been buried too deep for too long.

Pei Pei shook her head in disbelief.

"You really went through all of that alone?"

I nodded slowly.

"Back and forth to the police station, judicial department. Alone."

Abood looked down at his bowl for a moment, then said softly,

"You never showed it. But now I see it—those days you came in, barely lifting your head, avoiding eye contact... You weren't just tired. You were carrying something no one should carry alone."

I smiled faintly.

"I didn't even realize I looked that miserable. I thought I was doing a good job pretending."

"You were too good at it," Pei Pei said. *"But there were days we could feel the heaviness in the air when you walked in."*

We paused as the server refilled our drinks. I let the moment settle before I continued,

"It got worse when PIA came to the office that day, remember? Accusing me of having an affair with Abood."

Pei Pei scoffed.

"That man had no boundaries."

"I had to call the CEO and warn him,"

I said. *"I was shaking. I thought I was going to lose everything. But do you know what he told me?"*

They both looked at me, waiting.

"He said, 'I know who you are. As a woman, and as a professional. If your husband wants to talk, I'll meet him. I'll stand by you.'"

Pei Pei's eyes welled up a little.

"You deserved that. You deserved someone to say, 'I see you. I believe you.'"

We sat in a warm silence for a moment, letting the words and memories swirl between us. Then, just like that, Abood cracked a joke about how many chilies Pei Pei had thrown into her soup, and we burst into laughter—deep, belly laughter that felt like medicine.

That dinner was more than a meal.

It was a ritual of release.

For the first time in what felt like years, I wasn't explaining myself in fear or defending my actions. I was simply *seen*, heard, held in friendship, not judgment, and celebrated for surviving—not just the divorce but everything that led to it.

That night, as we stepped out into the cool night air, I looked up at the stars and whispered a silent *thank you*.

Not just for the freedom.

But for these two souls who reminded me that I was still human, still worthy, still here.

Chapter 19:

"Reclaiming the Girl I Left Behind"

A week after my 46th birthday, something unexpected arrived in my inbox: an offer letter from one of the top universities in Malaysia. I stared at it in disbelief, my fingers trembling slightly as I reread it to make sure I wasn't hallucinating.

I had applied months ago, on a whim, during one of my lowest points. That period was a blur of grief, stress, and countless back-and-forth trips handling my divorce. I remember sitting in my old studio, late at night, numb and restless, filling out the application form—not because I had hope, but because I needed something to anchor me. Something to distract me from the painful void that had taken over my life.

When the acceptance came, I felt a flicker of something I hadn't felt in a long time—purpose.

People often judged me for not having a degree. Some of them said it straight to my face, others whispered behind my back. It was easy for them to assume I wasn't capable, that I wasn't smart enough. But they never asked why. They never knew how I spent most of my life working and sending money home, sacrificing my own dreams so my family could live more comfortably. My dreams were buried quietly under the weight of responsibility.

I didn't lack intelligence. I lacked time, support, and privilege. That letter proved to me that it wasn't too late.

I enrolled immediately, fuelled by a mixture of spite and hope. I wanted to prove to those who had looked down on me that I was more than their assumptions. I wanted to prove to myself that I could still rise, even after being crushed emotionally, financially, and mentally.

When I received the acceptance call, I remembered the voice of the Dean who interviewed me. He had doubts. I could hear it in his tone when he questioned how I intended to manage university life—at my age, with such a basic entry score. His words weren't cruel, but they were cautious.

"I see that you've passed the minimum threshold," he said. *"But university life isn't easy. It's demanding, even more so for those returning after a long time."*

"I understand," I replied. *"But what I lack in academic background, I make up for in grit."*

He paused, then said, *"Well, we'll see."*

I took his hesitation as a challenge.

One of my favourite mantras is: *"What doesn't kill you makes you stronger."* Another one that's carried me through many storms is: *"If money can solve a problem, then it's not a real problem."*

By February 2021, COVID-19 was still raging globally. Most universities transitioned to online learning, which felt like a blessing to me. I could keep my full-time job in the UAE and still pursue my studies without having to relocate. I didn't need to resign or uproot my life again—I could finally do something for myself while still holding everything else together.

But I won't pretend it was easy.

After more than 30 years away from formal education, I felt like a fossil trying to navigate a digital maze. The online registration portal alone gave me anxiety. Terms like "student portal," "e-learning platform," "Turnitin," and "Zoom breakout room" felt like a foreign language.

I stared at the screen, overwhelmed, silently asking myself: *What have I gotten myself into?*

Luckily, my nephew was still studying at the same university. He guided me through the registration process, helped me figure out how to submit assignments, and showed me how to use Google Docs, shared drives, and formatting tools I'd never even heard of before. Within a couple of tries, I could manage the basics—but I was still terrified.

During my first week of classes, I made it a point to be honest with all my lecturers. In every subject, I introduced myself, not just by name, but by truth.

"Hi, I'm Liza. I've been away from school for 32 years. I'm good at operational work in the field, but I may struggle with academic writing and new technology. I ask for your patience and guidance."

I thought I'd be ignored or underestimated. But instead, I was met with kindness and support. Not once did any lecturer make me feel inferior. They helped me, one step at a time, with no judgment—just encouragement.

Still, my first assignment was rejected.

My lecturer returned it with a simple note: *"Redo. You can do better. Here's a guide."*

I could have crumbled right there. But something in his belief pushed me. I followed every piece of feedback he gave me, and slowly, I improved. Each assignment became a battle—and a small triumph. With every red mark, I learned something new.

But those victories weren't free. I paid for them with sleep, time, and isolation.

The time zone difference meant that my weekend classes started at 4:00 a.m. UAE time, which meant I had to be up by 3:00 a.m. every Saturday and Sunday.

On weekdays, I'd wake at 5:00 a.m. to work on assignments and leave for work at 7:00 a.m. sharp. I stopped going out, seeing friends, and living anything close to a "normal" life. My only companions were stress, caffeine, and the growing pressure to succeed.

Some nights, I cried out of sheer exhaustion. Some mornings, I questioned my choices.

But I never quit.

Sometimes, I'd sit behind the swabbing desk at the clinic, typing frantically on my laptop during breaks between patients. Sometimes, I'd hide in Abood or Pei Pei's office during an online class, praying no one would need me at that moment. They always supported me, never once complaining or making me feel like a burden. They became part of my academic journey, in ways they may never fully understand.

I can still remember one particular night. I was alone in the pantry, the hum of the fridge behind me and a blank document glowing on my screen. I had a paper due in six hours, and I was barely halfway done.

My hands trembled on the keyboard.

Why am I doing this to myself? I asked silently.

But then I thought of the girl I used to be—the one who always gave everything to others. The one who had dreams but shelved them. I owed her this.

I wasn't just chasing a degree. I was reclaiming a part of me I had lost.

The first year of my studies passed like a storm—chaotic, relentless, yet strangely quick. I was flying through it, not just surviving but excelling, carrying my books and bruises with equal grace. And just like that, I found myself at the one-year mark—not only in my academic pursuit but also in my divorce.

I didn't celebrate the anniversary. It wasn't something to commemorate, yet it lingered in the air like an old perfume—faint but unmistakable. Certain things still triggered me. The sharp chime of an iPhone, the clink of a bunch of keys, and the hum of a white sedan engine pulling into a driveway. They didn't paralyze me anymore, but they still whispered. Trauma doesn't leave all at once—it fades in layers. Maybe it helped that I'd moved out of that studio apartment. The silence there used to be unbearable, like it echoed every word he never said. My new place felt safer and brighter. Less haunted.

I don't remember exactly when I started talking about my divorce. It wasn't immediate. Maybe a year and a half after it was finalized. I just know it started one day in the car, during a casual chat with a good friend—the one I used to carpool with. She was the third person to know.

She blinked in surprise.

"You? Divorced? I never would've guessed."

I looked at her, half amused. "Why not?"

"You always seemed so composed. I thought maybe you were just... very private. But you didn't seem broken."

Of course, I didn't. I was trained to be invisible in my pain. At work, I kept my head down. I spoke to no men, kept to my three or four trusted friends, and stayed silent. Silence had always been my armour.

After that day, she checked in often—gently, without pressure.

"Do you need help with groceries?" "Want to talk?" "Just letting you know I'm here."

It meant more than I could explain. I told her not to tell anyone. I wasn't ready to share my story in public, and it wasn't the kind of story people wanted to hear anyway.

Most people didn't care—including my family.

None of my siblings ever asked if I was okay, if I needed food, or if I had a roof over my head. I was just their overseas ATM. The same siblings I helped send to university by sacrificing my own comfort would only message me when their wallets were empty. I guess that's what I became to them—a provider, not a person.

Maybe they think I'm strong. But strength doesn't mean I'm not human. I bleed in silence. I weep in the shower. I have a soul. I have feelings.

With my plate full—classes, assignments, work, bills—my life settled into an exhausting rhythm: wake up, study, work, sleep, repeat. Days blurred into nights. I didn't even realize when loneliness started creeping in. It wasn't loud. It crept in quietly, like a shadow during sunset.

Two years into the routine, the emptiness became too much. I needed a connection. Someone to talk to about my day. Someone who could understand the heaviness behind my smile. I didn't have time to go out and meet people—but my fingers, ever curious, found themselves exploring dating apps. Just to browse, I told myself.

But technology is like a trapdoor. One-click, and you're falling.

I stumbled upon a mankind, mature, and attentive. His messages were warm, his questions sincere. We graduated quickly from texts to video calls. He wasn't in the UAE anymore; he had just completed a project here and left. But distance didn't seem like a barrier. If anything, it made things feel more intimate. We talked daily. Morning check-ins. Nightly goodbyes. He became my routine.

Of course, when he asked to meet my family virtually, it felt like progress. COVID restrictions still made travel difficult, so online introductions seemed reasonable. He spoke of the future like it was already written—with me in it.

Looking back now, I see how masterful he was.

His life stories were full of struggle—loss, hardship, sacrifice. They tugged at my heart and made me want to believe him. I *did* believe him. There were a few odd things—tiny cracks in his stories—but he always had a convincing answer. And I, lonely and hopeful, chose to listen.

I thought I was finally being seen. But in reality, I was being studied.

The truth unravelled slowly, painfully. Promises turned vague. His needs began to surface—subtle financial requests cloaked as emergencies. That gnawing feeling in my gut returned, louder this time. When I finally connected the dots, the realization hit like a slap across the face.

He was a scammer—a smooth, well-rehearsed actor who preyed on career women like me—women with soft hearts and tired eyes.

And just like that, the love story I thought I had written became a cautionary tale.

Two years had passed since our first chat. In that time, we planned a wedding, postponed it, and re-planned it—only for each version to crumble under one excuse after another. Still, I clung to hope. Maybe because letting go would mean admitting I'd been living in an illusion. Or maybe because when you've been alone for too long, even a flicker of connection feels like sunlight.

Then one day, I did something bold. Something reckless. Something I thought was brave.

I resigned from my job.

Packed up my apartment.

Sold all my furniture and belongings.

And with five oversized suitcases, I boarded a plane to North Africa. He didn't know I was coming. I didn't want him to. I wanted to see what truth looked like when it wasn't staged.

When I arrived in Mascara, the air was thick with salt and dust. The city was foreign but not intimidating—like a dream, I wasn't sure I was supposed to be in. I checked into a modest hotel, the kind with sun-faded curtains and the scent of old wood varnish clinging to the walls. My heart pounded from travel fatigue and the sheer weight of what I was doing.

I texted him.

"I sent a courier to your address. They said you weren't home to receive it."

It was a lie. There was no courier, no package. I just wanted to know where he was, if he was even in the city, and if any part of his story still held up.

He replied almost immediately, confused.

"Why would you send something without telling me? What was it?"
"Why didn't the courier call me?"

There was hesitation in his tone. He knew something was off. A shift in energy. A disruption in the narrative he controlled.

And then I told him.

"I'm in Mascara. I came to see you."

There was a long pause. The kind that stretches so thin it could snap.

Then came his reply.

> "You've made my life a disaster."

That cut deep. I didn't expect gratitude or even joy—but not that.
He hung up.
But five minutes later, the phone rang again.
His voice had changed.

> "I'll take the next bus. I'm five hours away. Wait for me."

I didn't say much. I just nodded and ended the call, then collapsed on the hotel bed, still in my travel clothes. My shoes kicked off haphazardly. My body was aching from the long trip, but my chest ached even more—from uncertainty, from the cost of my decision, from all the things I gave up to be here.

He arrived that afternoon.

He looked different in person—more tired and less polished than he appeared on video calls. But still, something in his eyes felt familiar. We walked through the streets of Mascara in awkward silence. I remember the scent of grilled lamb in the air, the sound of traffic and merchants calling out in French and Arabic, and my heart thudding louder than all of it.

We went to a small cafe and ordered lunch we barely touched. The conversation was strained.

> "You should go back," he said quietly, not meeting my eyes.
> "You don't have a job anymore. There's nothing here for you."

I had already prepared for this. The disappointment. The deflection. I didn't come here without knowing it might fall apart. I had backup plans and a return ticket. But his words still shattered something inside me.

He promised to come for me, to pick me up from my hometown and bring me back to Mascara so we could "start fresh." But those promises now felt like feathers in the wind—beautiful in theory and weightless.

He didn't know that the version of me sitting across from him wasn't the same woman who once believed in fairytales. I was already broken inside. I wasn't looking for a happily-ever-after anymore. I just wanted the truth. I wanted closure. I wanted to face the man behind the voice and see if what we had was real.

And now that I had, I realised something painful:

I came all this way to find a man...

...only to find myself instead.

That afternoon in Mascara, we sat beneath a sun that felt too warm for the coldness between us. After our walk and a barely touched lunch, we sat on a bench near a roundabout where the city's noise blended with the heaviness in my chest. I asked him gently to explain himself.

He sighed, his voice barely above a whisper.

> *"Everything I've ever told you about my life was true. Nothing was fake. Nothing was a lie."*

I watched his eyes as he spoke. Searching for the flicker of truth I used to cling to.

> *"For the past few months... I didn't have a home. I was staying with my nephew and my brother. But my brother—he's deep into drugs. It's not safe. Not stable. I'm about to move to a new job. They'll give me shelter... finally."*

He looked away as if ashamed of the words leaving his mouth.

I sat there, frozen. Not because I lacked sympathy but because I had poured so much of myself into a man still trying to find his footing while letting me believe I was the missing piece. All this time, I thought I had seen the one—my companion until death, the man who would sit with me through the quiet of growing old.

He was polite. Soft-spoken. He carried knowledge about many things—philosophy, politics, even poetry. He knew how to listen, when to nod, what to say. His voice was calming, and his presence was always gentle. I convinced myself he was my soul's answer.

But now, sitting beside him in the real world—not through a screen or filtered words—I wondered if he'd simply mastered the art of pretending. Maybe he had rehearsed this part with others before me. Maybe I was just another name on his list and another woman moved by tragic stories and a polished charm.

"*He must have practiced this...*"

I whispered to myself later that night in the hotel room, the sound of car horns muffled through the closed window.

"*Practiced telling stories that made women feel needed. Then, once they care enough... once they're emotionally invested... he lets the sadness draw them in deeper. And maybe... they offer help. Money. Love. Loyalty. Whatever he needs.*"

Maybe I was just one of them.

I left Mascara with a suitcase heavier than when I came—not because of clothes, but the invisible weight of disillusionment. Still, I didn't let the pain swallow me. I returned with a different kind of fire.

Back in my country, I poured myself into finishing my studies. The coursework was getting more challenging each year, but it became my distraction. When the ache returned—when the silence from him lingered longer than it should—I buried myself in books, assignments, and lectures.

Each chapter, each assignment, each sleepless night became a plaster over my wounds. As the subjects got more complicated and more complex, so did my emotions—but at least one of them had a syllabus. At least one had an end date.

I replaced his absence with textbooks and lectures. I substituted what-ifs with research papers. There was no room for romance anymore—only results.

I told myself I would not fall again, not like that. Not for words spun too sweet, not for promises built on sand.

But even in the quietest moments—those few minutes before sleep or while stirring coffee in the early morning—I'd still hear his voice sometimes. Not in a haunting way. Just faintly. It's like a song that used to mean something.

And I'd ask myself: *was any of it real?*

Maybe not.

But my growth? My survival? My refusal to stay broken?

That was real—every inch of it.

I was jobless, yes. But not purposeless.

I started volunteering again—just like I did during the pandemic when the world ended when I tried to bring light to the darkest corners. This time, I collaborated with a local NGO. We ventured into remote villages, far from the city's shine. We delivered essentials—rice, oil, blankets. We played with the children and hosted simple parties that brought more laughter than money ever could. Some kids hadn't seen a balloon in months. Their eyes lit up at the smallest gestures.

Just being there—seeing them smile and their mothers smile—made me feel visible again like I mattered.

During the pandemic, it was more complex. We crossed borders with borrowed courage and fake sick leaves, sneaking past restrictions with the excuse of "urgent family matters." My brothers joined me, and together, we visited orphanages most people didn't even know existed. The air was always thick with sanitiser and fear, but laughter echoed like rebellion inside those houses.

Now, volunteering has become more than service. It has become survival. I may have been heartbroken, but at least I wasn't numb. At least I could still give love—even if I had no one left to give it to at home.

I wish I could have done more than volunteer. Sometimes, I would stand at the edge of those dusty village roads, watching the children scatter with balloons in their hands, the sun clinging to their skin like golden paint. I would close my eyes and imagine what it would be like to have the means to donate truckloads of supplies, to build a real school, to hand a mother a proper mattress instead of a thin foam pad.

But wishes don't stretch bank accounts.

I had already drained most of my savings—from the resignation, the failed trip, and the cost of trusting the wrong person. Being jobless didn't just mean no income—it meant I was watching my safety net unravel, thread by thread.

Still, I told myself: *You're not helpless. You're just starting again.*

I began searching for opportunities—not just jobs, but ways to reinvent myself. I was no longer the woman who once feared change; heartbreak and survival had carved a new kind of grit into me. I applied for roles that felt foreign, brainstormed side hustles late into the night, and even helped a friend run small-scale events to keep my mind occupied.

And then, in the thick heat of summer, I returned to the UAE.

I stepped off the plane and felt the hot wind slap my face, like a memory that refused to stay buried. The skyline of Abu Dhabi stretched before me like an old photograph—familiar but slightly faded. I stood still for a moment, clutching my luggage, watching people rush past, each with their own stories, none knowing how heavy mine had become.

"I thought this chapter was closed," I whispered as I entered a taxi. *"I thought I said goodbye for good."*

But life has a strange way of circling back—sometimes not as a punishment, but as an invitation to see things differently.
I wasn't the same woman who left.
I returned with bruises on my heart, but my eyes saw more clearly now. I no longer romanticised survival; I respected it. I knew that disappointment didn't always come with warning signs and that sometimes, faith meant taking the next step without knowing where the staircase led.
The driver asked, *"Where to?"*

"Just take me to the Corniche," I replied.
"I want to see the sea first."

As the city unfolded before me, I let my mind wander. Something about the sea permanently settled my spirit. The same waves I used to watch when I felt stuck now whispered something different: *You're back, but you're not broken. You're just beginning again.*
Maybe the Almighty had other plans— not built around the comfort of someone else's love but the strength of standing alone, rebuilding from the ground up, with nothing but grit, grace, and quiet prayer.
The salty breeze along the Corniche gave me just enough space to exhale for the first time in months. I sat on one of those low, sun-warmed benches, watching families laugh in the distance, cyclists zipping past, and birds dipping in and out of the water like they had no care in the world.
It wasn't peace exactly, but it was a pause. And in that moment, a pause was enough.
Within a few weeks, I started slowly rebuilding my work life. Familiar faces in the creative and retail industry greeted me with surprise and warmth. Some asked why I came back. I smiled and said, *"Sometimes life brings you full circle—not because you failed, but because you're ready to face it differently."*

I said yes to small projects, short contracts, and anything that gave me a sense of momentum. I threw myself into creating again—curating details, managing customer experience, and bringing designs to life. It was work I knew well, and though I carried invisible scars, I also brought new strength to the table.

But beneath the surface of the daily grind, I was chasing something far more personal—something no salary could give me.

I finished my studies.

Not just finished—*I excelled*.

I graduated with distinction. I walked onto that stage and received the best academic award, my name announced under bright lights and claps echoing across the hall. My certificate wasn't just paper. It was a declaration: *You tried to break me. You looked down on me. You left me out. And yet—here I am.*

> That scroll in my hand wasn't my future.
> It was my fight.
> It was proof that I was never less than anyone.
> Never.

Many thought I went back to school to change careers. They didn't know I went back to reclaim my dignity. To finish something I once abandoned. To give the little girl in me—who grew up constantly feeling invisible—the win she never thought she'd have.

And when I looked at my reflection that night, holding my award against my chest in silence, I whispered:

> *"You did it. You did it without him. Without anyone. Just you and the prayers no one heard but God."*

Final Chapter

"The Light Beyond the Wound"

Some say time heals all wounds. I don't fully believe that. Time doesn't erase the pain; it teaches you how to carry it gracefully. It shows you how to breathe through the weight, walk with your scars without shame, and smile even after the storm has shattered parts of your soul.

Today, my life isn't wrapped in ribbons or tied in perfect bows. It's not a fairytale. But I'm no longer lost. I'm no longer waiting to be rescued.

I'm learning. Slowly. Tenderly.

I'm still discovering who I am—this time, without shrinking to fit someone else's idea of love, loyalty, or worth.

I've allowed myself to love again, but not as I once did—not from a place of need or desperation.

Now, I love with softness and intention, with open eyes and steady hands—not to be saved, but because I have finally found the quiet courage... to be seen.

I've made peace with what I never received—support, protection, understanding—from those who once held pieces of my heart. I no longer bleed for people who never learned how to keep me gently.

I've stopped offering parts of myself to those who only ever took them without knowing how to give.

Now, I walk with those who stay.

Those who don't require noise to understand my silence.

Who feel my presence without needing an explanation.

Who sees me—not just the version I show the world, but all of me. Even the fragile, quiet corners I once kept hidden.

Looking back, I see her—

The woman who walked through fire.

Who didn't come out untouched... but came out *whole*.

Bruised, yes. But not broken.

Frightened at times but never truly defeated.

I see the nights she couldn't sleep—

On nights, she buried her screams into pillows,
And still got up the following day to face the world.
I see how she turned heartbreak into fuel,
And loneliness into wisdom.
And if you've made it this far into my story.
I don't know your name or what pain you carry,
But I know you came here hoping it's possible...
It is possible to survive.
To start again.
To rebuild something meaningful from the ashes.
And I want you to know:
It is.
This isn't the end of my story.
Life still surprises me. I still stumble. I still cry.
But now, I cry, and then I *rise*.
I fall, but I no longer stay down.
Because now,
I know when to speak... and when silence guards my peace.
I know when to hold on... and when to let go.
I know I don't need to be everything to everyone.
Most of all, I know this:
Even if my past held shadows,
Even if some chapters were written in pain—
There is still light.
There is still beauty left in this world.
And I will never stop seeking it.

Don't miss out!

Visit the website below and you can sign up to receive emails whenever QUEEN OF FLOWERS publishes a new book. There's no charge and no obligation.

https://books2read.com/r/B-A-LBYRD-EEVFG

BOOKS 2 READ

Connecting independent readers to independent writers.

About the Author

Queen of Flowers is a resilient soul whose journey spans cultures, heartbreaks, and spiritual growth. With a voice both tender and powerful, she shares her deeply personal story to inspire, heal, and remind readers that love, faith, and strength can bloom even through life's most testing storms.

www.ingramcontent.com/pod-product-compliance
Lightning Source LLC
Chambersburg PA
CBHW022342150525
26801CB00007B/101